Electronic and Nuclear Dynamics in Molecular Systems

Electronic and Nuclear Dynamics in Molecular Systems

Yuichi Fujimura
Tohoku University, Japan

Hirofumi Sakai
University of Tokyo, Japan

World Scientific

NEW JERSEY · LONDON · SINGAPORE · BEIJING · SHANGHAI · HONG KONG · TAIPEI · CHENNAI

Published by

World Scientific Publishing Co. Pte. Ltd.
5 Toh Tuck Link, Singapore 596224
USA office: 27 Warren Street, Suite 401-402, Hackensack, NJ 07601
UK office: 57 Shelton Street, Covent Garden, London WC2H 9HE

British Library Cataloguing-in-Publication Data
A catalogue record for this book is available from the British Library.

ELECTRONIC AND NUCLEAR DYNAMICS IN MOLECULAR SYSTEMS
Copyright © 2011 by World Scientific Publishing Co. Pte. Ltd.
All rights reserved. This book, or parts thereof, may not be reproduced in any form or by any means, electronic or mechanical, including photocopying, recording or any information storage and retrieval system now known or to be invented, without written permission from the Publisher.

For photocopying of material in this volume, please pay a copying fee through the Copyright Clearance Center, Inc., 222 Rosewood Drive, Danvers, MA 01923, USA. In this case permission to photocopy is not required from the publisher.

ISBN-13 978-981-283-722-6
ISBN-10 981-283-722-1

Typeset by Stallion Press
Email: enquiries@stallionpress.com

Printed by FuIsland Offset Printing (S) Pte Ltd. Singapore

Preface

Electronic and nuclear dynamics in molecular systems play a fundamental role in natural science and material science. The research targets are both physical and chemical phenomena occurring in simple diatomic molecules to complex biomolecular systems. In this monograph, we focus especially on quantum dynamic behaviors of molecular systems in which electronic and/or nuclear motions driven by laser pulses are treated quantum mechanically. In the 1980's to 2000, *nuclear quantum dynamics* in molecules was extensively investigated by many researchers including the Nobel Prize winner Ahmed Zewail, the pioneer of femtosecond chemistry. One of the greatest achievements in nuclear quantum dynamics is real-time observation of transition states in the course of chemical reactions. This was made possible by the use of ultra-short pulses in femtosecond time regimes and their detection system. This is now called "femtosecond chemistry". In the first decade of the twenty-first century, there have been further developments in laser science and technology: pulse-shaping technique, control of carrier-envelope phase and attosecond pulse generation. These developments have made it possible to directly observe and control motions of electrons in molecules, although the molecules are restricted to simple molecules at the present time. The results have provided fundamental ideas for realization of functional devices such as ultrafast switching devices and quantum computers. Therefore, it is timely to overview recent developments in electronic and nuclear quantum dynamics of molecular systems from both experimental and theoretical points of view.

The purpose of this monograph is not only to review quantum dynamical behaviors of electrons and nuclei of molecular systems but also to introduce experimental and theoretical methods for graduate students and/or beginners who have become interested in this field. Gaps between the levels of beginners and researchers have become larger and larger. The authors hope that this monograph will contribute to bridging those gaps.

This monograph is organized into two parts and eight chapters; in the first part, i.e., Chap. 1, fundamental concepts and principles of coherent dynamics of electrons and nuclei are described. These include wavepackets, adiabatic and nonadiabatic treatments, quantum optimal control, and multiphoton vs. tunneling ionization. In Chap. 2, the experimental apparatus and methods for measuring ultrafast dynamics are presented.

In the second part, Chaps. 3 to 8, the results of experimental and theoretical studies on electronic and nuclear coherent motions are presented. Methods for time evolution of electronic and nuclear wavepackets created by laser pulses are described in Chap. 3. Typical numerical methods are given for solving the time-dependent Schrödinger equation of molecular systems. Here, the semi-classical treatment for interactions between molecules and lasers is adopted on the basis of the assumption that quantum behaviors such as spontaneous emission can be neglected. Chapter 4 outlines experimental setups and techniques for molecular alignment and orientation with laser technologies first. Then, various applications with a sample of aligned or oriented molecules are reviewed to prove their usefulness. Alignment and orientation of molecules are essential processes for the studies of selective chemistry under the conditions of vapor and condensed phases. In Chap. 5 theoretical treatments of electronic and nuclear dynamics in intense laser fields are shown. The correlated motion between the electronic and nuclei in a hydrogen molecular ion is presented. Interelectronic correlations in the simplest two-electron system, a hydrogen molecule, are shown. Nonlinear dynamics, Coulomb explosion, multiple ionization and decomposition of fullerenes induced by intense laser pulses are presented from both experimental and theoretical points of view. Chapter 6 presents theoretical studies on π-electron rotations of molecules, Mg porphyrin, benzene and ansa (planar chirality) chiral aromatic molecules, which are induced by ultrashort laser pulses. Chapter 7 shows real-time observation of stilbene cis-trans coherent isomerization in condensed phases, which is one of the typical examples of photochemical reactions. Results of experiments on laser control of retinal isomerization in rhodopsin are also shown. Chapter 8 presents applications of quantum control theory to molecular chirality. Several methods such as pump-dump control, stimulated Raman adiabatic passage, and quantum control methods are applied. Molecular chirality is an essential property of biomolecules. The development of new methods for identification and transformation of chirality is a challenging subject.

We would like to thank our colleagues, co-researches and friends: Profs. S. Aono, A. D. Bandrauk, S. L. Chin, P. B. Corkum, W. Domcke, L. González, I. V. Hertel, M. Kanno, S. Koseki, Z. Lan , F. H. M. Faisal, G. Gerber, S. H. Lin, R. J. Gordon, M. Hayashi, K. Hoki, T. Kato, N. Kobayashi, H. Kono, D. Kröener, Y. Manz, D. Mathur, K. Miyazaki, H. Nakamura, Y. Nomura, T. Nguyen-Dan, Y. Ohta, Y. Ohtsuki, R. Sahnoun, N. Shimakura, M. Sugawara, M. Tanaka, K. Terarishi, H. Umeda, K. Yamanouchi, K. Yamashita and Y. Yan, Drs. I. Kawata, K. Nakai, Y. Sato and M. Yamaki, Ms. M. Abe, and K. Haruyama, A. Kita, M. Muramatsu and

K. Nakagami. YF would like to express special thanks to Prof. H. Kono and Prof. Y. Ohtsuki who were partners in carrying out hard-time research projects at Tohoku University, Sendai Japan. Professor H. Kono has made a substantial contribution to the development of electronic quantum dynamics with Dr. I. Kawata.

Parts of this monograph were completed while YF was a research fellow at the Institute of Atomic and Molecular Sciences with support by Academic Simica and National Science Council of Taiwan, to which Prof. Lin kindly invited him. YF appreciates the warm hospitality of M. Hayashi, Professor of Center for Condensed Matter Physics.

The staff at World Scientific Publishing were exceedingly helpful, including Editors V. K. Sanjeed and C. Xingliang.

This monograph is dedicated to the late Professors T. Nakajima and T. Abe, who provided support and encouragement at the initial stages of YF's research projects.

The authors acknowledge financial support from JSPS (Japan Society for the Promotion of Science) and MEXT (Ministry of Education, Culture, Sports, Science, and Technology).

Contents

Preface v

1. Basic Concepts and Fundamental Dynamics 1
   1. Wavepacket 1
   2. Adiabatic and Nonadiabatic Treatments 3
   3. Quantum Control by Pulse Shaping 4
   4. Multiphoton vs. Tunneling Ionization 6
   5. High-Order Harmonic Generation (HHG) 7
   6. Rescattering of Electrons 10
   7. Above-Threshold Ionization (ATI) 11
   8. Above-Threshold Dissociation (ATD) 12
   9. Coulomb Explosion 13
   10. Alignment and Orientation 15
   11. Molecular Chirality 17
   12. Molecular ADK Theory for Tunneling Ionization Rate 19
   13. Summary 21

2. Experimental Setups and Methods 25
   1. Laser-Induced Coulomb Explosion Imaging 25
   2. High-Order Harmonic Generation 26
   3. Molecular Orientation and Its Observation 28
      3.1 1D molecular orientation 28
      3.2 3D molecular orientation 30

3. Theoretical Treatments of Wavepackets 35
   1. Generation of Wavepacket and Its Propagation 35
   2. Numerical Methods for Wavepacket Propagation 39

		2.1	Symmetrized split operator method	40
		2.2	Finite difference methods	41
		2.3	Dual transformation technique	45
	3.	Wavepacket Propagation Method in the Scattering Matrix Framework		49

4. **Molecular Manipulation Techniques with Laser Technologies and Their Applications** 55

1. Introduction 55
2. Techniques for Molecular Alignment 56
   - 2.1 Adiabatic and nonadiabatic alignment 56
   - 2.2 1D and 3D alignments 58
3. Techniques for Molecular Orientation 59
   - 3.1 1D and 3D orientations 59
   - 3.2 Laser-field-free molecular orientation 59
   - 3.3 All-optical molecular orientation 60
4. Various Applications with a Sample of Aligned or Oriented Molecules 64
   - 4.1 Tunnel ionization 64
   - 4.2 Nonsequential double ionization 65
   - 4.3 Control of multiphoton ionization 66
   - 4.4 Control of photodissociation 67
   - 4.5 High-order harmonic generation 68
   - 4.6 Structural deformation of polyatomic molecules 71
   - 4.7 Selective preparation of one of the enantiomers 71
   - 4.8 Phase effects and attosecond science 73
   - 4.9 Polarizability anisotropies of rare gas van der Waals dimers 73
   - 4.10 Probing molecular structures and dynamics by observing photoelectron angular distributions from aligned molecules 75
   - 4.11 Observation of photoelectron angular distributions from oriented molecules with circularly polarized femtosecond pulses 76
5. Concluding Remarks 77

5. **Electronic and Nuclear Dynamics in Intense Laser Fields** 81

1. Electron–Nuclei Correlated Motions in H_2^+ 81
   - 1.1 Hamiltonian in terms of cylindrical coordinates 82
   - 1.2 Phase-adiabatic states 85
   - 1.3 Interwell electron transfer 87
   - 1.4 Dissociative ionization 93

	2.	Interelectronic Correlation in a Hydrogen Molecule	95
		2.1 Hamiltonian of H_2 in the presence of laser fields	95
		2.2 Electronic wavepacket dynamics of H_2	97
	3.	Reaction Dynamics of Carbon Dioxide	99
		3.1 Doorway states for Coulomb explosions	99
		3.2 Cycle-averaged potential energy surface and simultaneous two-bond breaking	100
	4.	Benzene	104
	5.	Fullerene	108
	6.	Ejection of Triatomic Hydrogen Molecular Ions from Hydrocarbons	112
6.	Electron Rotation Induced by Laser Pulses		117
	1.	Introduction	117
	2.	Electronic Ring Currents Generated by Circularly Polarized Laser Pulses	118
		2.1 Magnesium-porphyrin	118
		2.2 Benzene	120
	3.	Control of Unidirectional Rotations of π-Electrons in Chiral Aromatic Molecules	122
	4.	Nonadiabatic Effects of Laser-Induced π-Electron Rotation	127
7.	Photoisomerization and Its Control		133
	1.	Introduction	133
	2.	Real-Time Observation of Stilbene *Cis–Trans* Isomerization	134
	3.	Quantum Control of Retinal Isomerization in Bacteriorhodopsin	137
	4.	Quantum Control of Retinal Isomerization in Rhodopsin	139
8.	Quantum Control of Molecular Chirality		147
	1.	Molecular Chirality Transformation in a Preoriented Racemic Mixture	147
	2.	Pump–Dump Control *via* an Electronic Excited State	154
	3.	Stimulated Raman Adiabatic Passage Method	159
	4.	Control of Helical Chirality	163
	5.	Quantum Control in a Randomly Oriented Racemic Mixture Using Three Polarization Components of the Electric Fields	167
Index			173

Chapter 1

Basic Concepts and Fundamental Dynamics

In this chapter, basic concepts and fundamental dynamics treated in this monograph are introduced for the nonexperts. The concepts introduced include wavepackets, adiabatic and nonadiabatic treatments, quantum optimal control, multiphoton vs. tunneling ionization, rescattering of electrons, and molecular chirality. High-order harmonic generation (HHG), above-threshold ionization (ATI), above-threshold dissociation (ATD), and Coulomb explosion are introduced as fundamental dynamics in intense laser fields.

1. Wavepacket

Wavepacket is a nonstationary and localized state of atoms or molecules.[1] There are various types of wavepackets, such as rotational wavepacket, nuclear wavepacket, Rydberg wavepacket, and electronic wavepacket, depending on the degrees of freedom of the dynamics. The use of wavepackets has played a central role in the development of femtosecond chemistry.[2] The concept of a wavepacket is important for application of wavepacket methodology such as optical science as well as molecular science.[3]

In most cases, due to different characteristic time scales in these dynamics, each wavepacket is assumed to be independent from the others. For example, for treating nuclear wavepackets of a molecular system, vibrational degrees of freedom are taken into account, while rotational degrees of freedom are assumed to be frozen; the electronic wavepacket is washed away within the time scale of creating a vibrational wavepacket.

To understand qualitatively how wavepackets are created by ultrashort-pulsed lasers and how they propagate on the potential energy surface, consider a one-dimensional (1D) and vibrational motion within a two-electronic state model as shown in Fig. 1. The initial state is taken as the lowest vibrational eigenstate in the electronic ground state, g, X_{g0}. We assume a delta-function excitation, omitting effects of the pulse excitation. $X_e(0)$ is the excited-state wavepacket created from

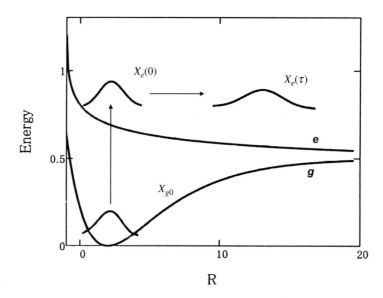

Fig. 1. A schematic illustration of creation of a nuclear wavepacket and the time evolution in a two-electronic state model. The arrows indicate the temporal evolution of the system. The lower curve denotes the potential energy curve in the ground state g and the upper curve indicates that in an electronic excited state e. X_{g0} is the lowest vibrational eigenstate in the ground electronic state, $X_e(0)$ is the excited-state wavepacket created from X_{g0} by a pulsed laser field, which is derived based on a perturbative treatment, and $X_e(\tau)$ is the wavepacket on the excited-state potential at time τ.

X_{g0}, and $X_e(\tau)$ is the wavepacket at time τ. Within a constant electronic transition moment, $X_e(0)$ has the same form as that of X_{g0}. The nuclear wavepacket localized on the excited-state potential surface spreads, delocalizes, convergences, and localizes again with elapse of time. The dynamical behavior is repeated. Such nuclear dynamics can be explained in terms of interferences between the eigenstates in the electronic excited state.

A wavepacket $X_e(\tau)$ can be described in terms of a linear combination of the vibrational eigenstates in the excited state. That is, a coherent excitation by pulses creates a linear combination of the eigenstates. The localized wavepacket changes its form by interferences between the eigenstates as time evolves and becomes delocalized. The process of losing localization is called dephasing. For low-dimensional systems, such as a one- or two-mode system, a wavepacket that has been delocalized becomes localized again. This process is called rephasing or revival.[4] This is one of the typical and quantum mechanical phenomena.

The creation of a localized wavepacket and its time evolution can be directly observed by time-resolved spectroscopic methods such as time-resolved fluorescence and ionization methods.[5] In the time-resolved spectra, oscillations called quantum beats appear because of interference between eigenstates that are coherently excited.[6]

In recent years, the dissociative ionizations of molecules in intense laser fields have been observed.[7] The dynamics originates from electron-nucleus-correlated motions. In order to clarify the correlated dynamics, it is necessary to take into account the electronic and nuclear wavepackets in which the electronic and nuclear degrees of freedom are variable. The electronic and vibrational dynamics of molecules in terms of electronic and nuclear wavepackets will be described in detail in Chap. 5.

2. Adiabatic and Nonadiabatic Treatments

Large differences by the thousand in mass ratios between electrons and nuclei make it possible to separate electronic and nuclear (vibrational) motions in molecules.[8] In other words, electrons adiabatically follow nuclear motions. This separation is called adiabatic or Born–Oppenheimer approximation, and it is the starting point in understanding the dynamical behaviors, especially in the ground electronic state of molecules. The separation breaks down when the potential energy surfaces of two electronic states cross each other. In this case, nonadiabatic effects become important in treating electronic and nuclear dynamics. The excited electronic energy is transferred into energies of nuclear motions of the nuclear kinetic energy.[9] Typical nonadiabatic effects in photochemical reactions of aromatic biomolecules are described in Chap. 7.

When molecules are placed in strong laser fields, electrons in molecules follow the oscillating electric field and move on the time-dependent electron potential (electrostatic potential + molecule-laser field interaction) for certain times. Such a molecular state is called the phase (field) adiabatic state. One of the electron dynamics that takes place in the phase adiabatic state can be seen in electron transfers between localized states. As the distance between the two localized states stretches, the ability of the electron transfer decreases because nonadiabatic interactions between the two adiabatic states make a significant contribution. Theoretical treatments based on the phase adiabatic states are presented in Chap. 5.

Adiabatic and nonadiabatic treatments can be seen in control of molecular alignments and orientations as well. In the adiabatic treatment, the control procedure is carried out slowly compared to the rotational period of molecules. In the nonadiabatic treatment, control is carried out within the time regime in which the molecules are in a nonstationary state. Experimental results of alignment and orientation of diatomic and small polyatomic molecules and their analysis will be shown in Chap. 4.

3. Quantum Control by Pulse Shaping

In general, quantum control means the manipulation of the target of interest at the quantum level.[10] There are two types of quantum control: one is called coherent phase control and the other is called pulse shaping. The first type utilizes the interferences between two optical quantum processes in a constructive or destructive way by changing the phase of an applied stationary laser and the latter type manipulates wavepackets by tailored laser pulses. There have been many applications of quantum control to chemical reaction dynamics.[11] In this section, we focus on quantum control by pulse shaping.

Figure 2 shows pulse shaping using a liquid crystal as a spectral modulator.[12] A transformed-limited pulse is introduced into a grating mirror (G_1) that is placed at the focal point of the lens (L_1). A spatial light modulator (SLM) is placed in the Fourier plane to alter the phases and/or amplitudes of the light in each frequency component. For this purpose, a special liquid-crystal light modulator and acoustic-optic modulator are used. The modified pulse is finally recompressed by the lens (L_2) and grating (G_2).

In most cases, the molecular Hamiltonian and interactions with laser fields cannot be explicitly determined. There exist uncertainties in the pulse-shaping system. To avoid these difficulties, a genetic algorithm (GA) is usually adopted. In a GA procedure, three operations, *i.e.*, reproduction, crossover, and mutation, are carried out to obtain the best results. Thus, the GA procedure has a feedback loop structure without any involvement by experimentalists. An example of a GA experiment is the

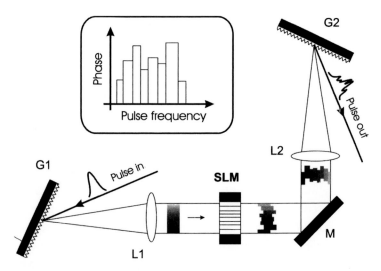

Fig. 2. A scheme for pulse shaping using a liquid-crystal modulator. (Reprinted with permission from Uberna *et al.*[12] Copyright (1999) by the Royal Society of Chemistry).

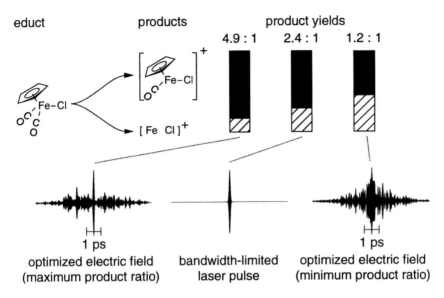

Fig. 3. Application of a genetic algorithm to the control of the photodissociative ionization of cyclopentadienyl-iron-dicarbonyl-chloride $C_2H_5Fe(CO)_2Cl$. Electric fields of the laser generated to either maximize or minimize the ratios of the products $C_2H_5FeCOCl^+$ to $FeCl^+$ are shown at the bottom. Provided by the courtesy of T. Brixner, and adapted from Assion et al.[13]

control of dissociative ionization products of $C_2H_5Fe(CO)_2Cl$ as shown in Fig. 3.[13] To maximize or minimize the ratios of the photo-cyclopentadienyl-iron-dicarbonyl-chloride $C_2H_5Fe(CO)_2Cl$ was the control target. Figure 3 shows the results of the optimum ratio and the pulse shapes obtained by the GA.

Theoretically, the optimal control field $E(t)$ is designed to maximize the objective functional $O[E(t)]$ as[14]

$$O[E(t)] \equiv <\Psi(t_f)|W|\Psi(t_f)> - \int_0^{t_f} \frac{dt}{\hbar A(t)}|E(t)|^2,$$

where W is a target operator, t_f is a control time, and $A(t)$ is a regulation function for tuning laser intensity. $\Psi(t)$ is the wavefunction that satisfies the time-dependent Schrödinger equation of the system with an initial condition. Using the variational principle, $\delta O = 0$, an optimal electric field $E(t)$ is obtained as

$$E(t) = -A(t)\text{Im} <\Xi(t)|\mu|\Psi(t)>.$$

Here, $\Xi(t)$ is the wavefunction that satisfies the time-dependent Schrödinger equation of the system with the final condition and μ is the dipole moment operator.

In the latter chapters, the application of the optimal control to photochemical reaction dynamics involving nonadiabatic coupling effects will be described.

4. Multiphoton vs. Tunneling Ionization

Ionization is the simplest and best understood electron dynamics of atoms and molecules at least in weak laser fields. As the intensities of lasers increase, the ionization processes become more complicated since the electronic potential is subjected to change. Let us now consider ionization processes of a hydrogen atom by an intense IR laser field.

Figure 4 shows a 1D model of a hydrogen atom in a linearly polarized electric field applied to the direction of the electron motion. The electron moves back and forth in the oscillating electric field of the laser. There are two pictures to describe the ionization processes. The first one is multiphoton ionization since a number of IR photons are necessary. The other picture is tunneling (field) ionization since an intense laser field in the long wavelength limit can induce ionization by tunneling through the distorted electronic potential as shown in Fig. 4(b).

In the former case, ionization rates can be evaluated by the perturbation method, while in the latter case, ionization rates can be evaluated by the nonperturbative methods to solve directly the Schrödinger equation of electrons in intense laser fields. The criterion by which the picture of ionization is more appropriate to describe ionizations is given by parameter γ, so-called Keldysh parameter: $\gamma = \sqrt{I_p/(2U_p)}$, where I_p is the ionization potential and U_p is the pondermotive energy that is the time-averaged energy of an electron in the oscillatory electromagnetic field.

The parameter γ was introduced by Keldysh in deriving the transition probability of one-electron atoms from an initial bound state to a Volkov state, *i.e.*, exact electronic state in a field in the low-frequency limit.[15]

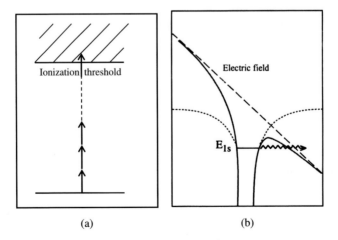

Fig. 4. Two ionization processes: (a) multiphoton ionization and (b) tunneling ionization. A 1D motion of the electron in a hydrogen atom is taken as a model for tunneling ionization.

It can be shown by using the above simple model that the Keldysh parameter is expressed in terms of the ratio between electron tunneling time τ_{tun} and half of the cycle τ_L of the laser used as $\gamma = \tau_{\text{tun}}/(\tau_L/2)$. This relation can be obtained by using two expressions: one is the pondermotive energy given as $U_p = E_0^2/(4\omega^2)$ with the electric field amplitude E_0 and laser frequency ω, and the other is the tunneling time $\tau_{\text{tun}} = l/v_{\text{el}}$ with the length of barrier l through which the electron tunnels and the velocity of the electron v_{el}. We use two auxiliary relations. One is the length of the barrier that is obtained by using the expression $I_p = E_0 l$ assuming that the electron ionizes from the lowest 1s state with energy at $-I_p$, ionization potential, and is ejected with zero kinetic energy. The other is the relation between v_{el} and I_p before the laser field is switched on, $I_p = m_e v_{\text{el}}^2/2$ or $v_{\text{el}} = \sqrt{2I_p}$ in atomic units.

From the Keldysh parameter, the tunneling ionization picture is appropriate to explain ionization processes under the condition $\gamma < 1$, *i.e.*, the electron tunneling time is short compared with the oscillation periods.[16] On the other hand, the multiphoton ionization picture is appropriate for $\gamma \geq 1$.

5. High-Order Harmonic Generation (HHG)

High-order harmonic generation (HHG) is the best known electronic dynamics in intense laser fields.[17] In HHG processes, the energies of multiphoton of the incident IR laser field are transformed into a single photon energy in UV regime above the ionization threshold. Ionization is an incoherent process between matter and the radiation field, while HHG is a coherent nonlinear optical process similar to second harmonic generation (SHG), sum-frequency generation (SFG), *etc.* as shown in Fig. 5(a). Coherent interactions of atoms or molecules with an intense laser field create more than 10th orders of harmonic generation. High-order harmonic generations with extreme ultraviolet (XUV) or x-ray wavelengths, which are created by using femtosecond IR or VIS laser pulses, are expected to be an x-ray source for a real-time measurement of molecular structures in vapor or condensed phases.[18]

High-order harmonic generation has extensively been studied in rare gases since 1987.[19] High-order harmonic generation spectra of rare gas atoms He and Ne, created by intense subpicosecond dye laser pulses, are shown in Fig. 5(b).[20] An HHG spectrum is characterized by a region with constant energy that is called a plateau and a point where a sudden decrease starts, which is called the cutoff. For He, the cutoff appears around 36 harmonic order. In Fig. 5(b), odd orders of the peaks appear in the HHG spectra. This is called the dipole selection rule for HHG. This selection rule is applicable to HHG from atoms and molecules with inversion symmetry.

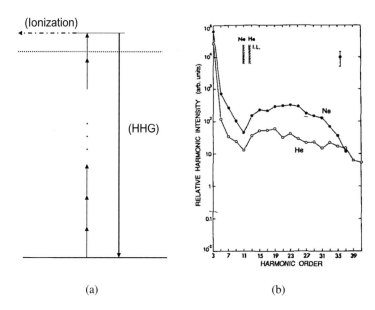

Fig. 5. (a) Scheme of high-order harmonic generation (HHG) that is one of the coherent nonlinear optical processes. For comparison, ionization that is an incoherent process is shown as well. (b) HHG spectra of rare gases, helium, and neon atoms, which are generated by subpicosecond dye laser pulses. (Reprinted with permission from Miyazaki and Sakai.[20] Copyright (1992) by Institute of Physics).

The first experiment on HHG from molecules, diatomic molecular gases, was reported in 1994.[21] Theoretical works on molecular HHG were carried out on H_2[22] and H_2^+,[23] preceding the experiment. It has been shown that molecular polarizability is an important factor for HHG efficiencies rather than the ionization potential and mass.

It has been experimentally shown that the HHG spectra from diatomic molecules and polyatomic molecules are characterized by plateau and cutoff similar to atomic HHG spectra, as shown Fig. 6.[24] Molecules have rotational and vibrational degrees of freedom, not present in atoms. Therefore, it is interesting to investigate alignment and photon polarization effects and other effects characteristic to molecular HHG.[25]

The mechanism of HHG is simply explained in terms of rescattering of an electron in the quasi-static rescattering model within classical mechanics.[26] In this model, electronic potential effects are neglected since the pondermotive energy is much larger than the ionization potential. In the rescattering model, HHG is divided into three steps, *i.e.*, (1) tunneling ionization, (2) energy-gain, and (3) recombination, as shown in Fig. 7. In the first step, an electron in the bound state tunnels through the potential distorted by the intense laser field and being ejected from the bound state to a continuum. In the second step, the ejected electron gains highly kinetic energy from the laser field. In the final step, recombination step, a high-energy photon is

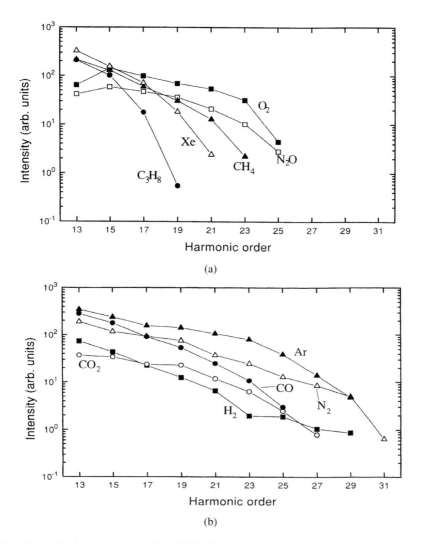

Fig. 6. High-order harmonic generation (HHG) from molecules. For comparison, HHG from atoms are shown as well: (a) HHG from atoms and molecules with a low ionization potential (from 10.9 to 12.9 eV) and (b) those with a higher ionization potential (from 13.8 to 15.8 eV). (Reprinted with permission from Lyngå et al.[24] Copyright (1996) by Institute of Physics).

emitted when the electron returns to the initial state after the oscillating field changes the sign.

A linearly polarized laser is necessary for rescattering of an electron with its core. In a circularly polarized laser field, the ionized electron cannot return back to its ion core because of the angular momentum of photon transfers to the electron.

The maximum energy in HHG created from atoms is known as $3.17 U_p$, called the cutoff law. The maximum velocity of an electron passing the ion core corresponds to

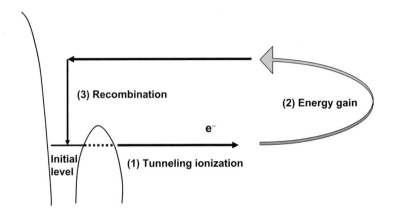

Fig. 7. Three-step model for HHG based on the quasi-static rescattering scheme: the first step is the electron tunneling, the second step is energy gain of the electron from intense electric field of laser, and the third step is the recombination of electron to the initial electronic state.

an instantaneous kinetic energy of 3.17 times the pondermotive energy. The cutoff law can be derived in the rescattering model. In molecular HHG spectra, a deviation from the cutoff law is observed because of effects of vibrations of the ion core.[27]

6. Rescattering of Electrons

In Sec. 4, we surveyed the tunneling of an electron in an intense field from a static viewpoint. In this section, we consider the behaviors of an electron in an intense oscillating field of a laser from dynamical viewpoints. The electron ionized by tunnel ionization comes back to its ion core in a linearly polarized intense oscillating field. The electron is then scattered by the ion core. These processes are repeated by the oscillating field. The dynamical behavior is called rescattering. The rescattering is the origin of high-order harmonic generation (HHG) as described in Sec. 5.

Let us now estimate the cutoff energy (the highest energy) in HHG spectra of an atom using the rescattering of an electron in the classical model.[26] The equation of motion of an electron in an oscillating electric field of a laser, $E(t) = E_0 \cos(\omega t + \varphi)$, is expressed as (in atomic units)

$$\frac{d^2}{dt^2} r(t) = -E_0 \cos(\omega t + \varphi), \tag{1}$$

where E_0 is the maximum electric field of frequency ω and the initial phase φ.

$$\frac{d}{dt} r(t) = -\frac{E_0}{\omega} [\sin(\omega t + \varphi) - \sin \varphi], \tag{2}$$

and

$$r(t) = \frac{E_0}{\omega^2} [\cos(\omega t + \varphi) - \cos \varphi + \omega t \sin \varphi]. \tag{3}$$

The return of the electron to its ion core is called recombination. Recombination of an electron having maximum returning velocity at $r = 0$ creates the maximum energy in the HHG signal. We note from Eq. (3) that at $r = 0$ the phase has to be satisfied with condition,

$$\tan \varphi = \frac{\cos(\omega t) - 1}{\sin(\omega t) - \omega t}. \tag{4}$$

The optimal condition for maximum velocity at $r = 0$ is expressed using Eq. (2) as

$$\frac{d}{dt}\left(\frac{dr}{dt}\right) = -\frac{E}{\omega}\left[\cos(\omega t + \varphi)\left(\omega + \frac{d\varphi}{dt}\right) - \cos\varphi \frac{d\varphi}{dt}\right] = 0. \tag{5}$$

Solving Eq. (5) with Eq. (4), we obtain

$$2(1 - \omega t \sin \omega t) + ((\omega t)^2 - 2)\cos \omega t = 0. \tag{6}$$

One of the solutions is given by $\omega t = 1.30\pi$ and $\varphi = 0.13\pi$ in a good approximation. The time for the electron to arrive at the origin is estimated to be $t = 1.7$ fs for an accelerating laser with wavelength $\lambda = 800$ nm.

The energy E_{max} at the peak of the HHG signal is then expressed as

$$E_{max} = \frac{1}{2}\left(\frac{dr}{dt}\right)^2 = \frac{1}{2}\left(1.26\frac{E_0}{\omega}\right)^2 = 3.17 U_p. \tag{7}$$

Here, U_p is called the pondermotive energy. This is the time-averaged energy of an electron in the oscillatory electromagnetic field and is given as $U_p = \frac{1}{4}\alpha_p^2 \omega^2$, where $\alpha_p = \frac{E_0}{\omega^2}$ is called the pondermotive radius.[7]

It should be noted that the cutoff energy derived above is valid for an electron under a Coulomb interaction-free condition.

As shown in Fig. 5, this value explains the observed signal peaks of HHG from atoms. This is called the $3.17 U_p$ law for the HHG peak as well.

7. Above-Threshold Ionization (ATI)

When an atom or molecule is ionized by an intense laser pulse, more photons than the minimum number needed to reach the ionization threshold are absorbed.[28] The ionization process involving free-free transitions is called above-threshold ionization. The first observation of ATI was free-free transitions following six-photon ionization of xenon atoms by Nd^{3+}-glass laser ($\hbar\omega = 1.17$ eV).[29] The intensity was about 4×10^{13} W/cm^2 with a pulse duration of 12 ns. Seven years later after observation of the atomic ATI, experimental results on ATI of molecular hydrogen were reported.[30] The ATI process through its resonant multiphoton ionization was observed. Figure 8(a) shows the photoelectron spectra following the five- and six-photon (one additional photon above the ionization potential) of H_2 via the Q(1) four-photon transition (E, F $^1\Sigma_g^+$, v_E) ← ($X^1\Sigma_g^+$, $v = 0$).[31] The schematic energy-level diagram relevant to the photoelectron spectra of H_2 is shown in Fig. 8(b).

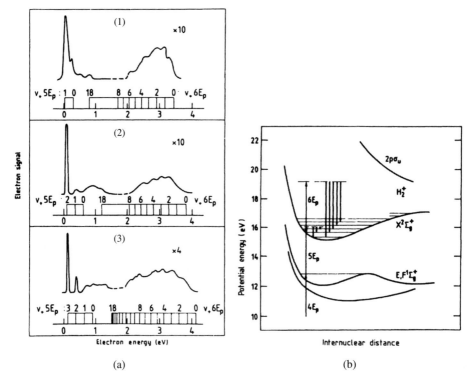

Fig. 8. (a) Photoelectron spectra following five- and six-photon (one additional photon above the ionization potential) ionizations of H_2 via the Q(1) four-photon transition (E, $F\,^1\Sigma_g^+$, v_E) ← ($X^1\Sigma_g^+$, $v = 0$): (1) $v_E = 1$; (2) $v_E = 2$; and (3) $v_E = 3$. (b) Schematic energy-level diagram relevant to the photoelectron spectra. (Reprinted with permission from Cornaggia et al.[31] Copyright (1986) by the American Physical Society). See Fig. 9(b) for the energy-level diagram of H_2 including the ground state potential.

8. Above-Threshold Dissociation (ATD)

A molecule has continuum states of the nuclei as well as continuum states of the electrons. Free-free transitions in the vibrational continuum states are observed in a molecule together with ATI under an intense laser field condition.[32] The free-free transition in the vibrational continuum states is called above-threshold dissociation (ATD). The mechanisms of ATD have been investigated experimentally and theoretically.[33]

Figure 9 shows an ATD spectrum of a hydrogen molecule: signals of a hydrogen molecular ion and hydrogen ion produced via six-photon ionization of H_2 through four-photon resonance on E, $F^1{}_g{}^+ v_E = 0$ (vibrationless state) are plotted as a function of the laser wavelength at laser intensity of 1.5×10^{11} W/cm^2.[34] The energy level diagram relevant to the ATD process is given in Fig. 9(b) for understanding the ATD mechanism.

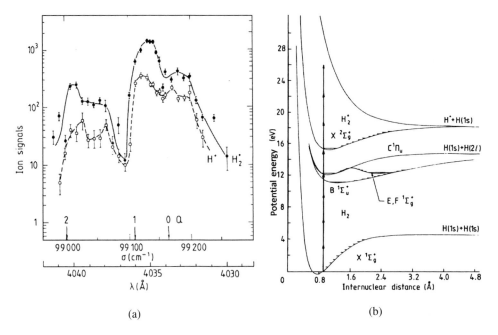

Fig. 9. (a) Above-threshold dissociation (ATD) spectrum of a hydrogen molecule. (b) Potential energy diagram of H_2, which is associated with the ATD. (Reprinted with permission from Normand et al.[34] Copyright (1986) by Institute of Physics).

The fragment ions H^+ are formed by photodissociation of H_2^+ after six-photon absorption as

$$H_2(X, v = 0) + 6h\nu \rightarrow H_2^+(v_+) + e^-,$$
$$H_2^+(v_+) + h\nu \rightarrow H^+ + H(1s).$$

9. Coulomb Explosion

Coulomb explosion is one of the dissociative ionizations of molecules induced by intense laser fields.[35] For a diatomic molecule AB, for example, multicharged and dissociative fragment ions, A^{m+} and B^{n+}, are produced after Coulomb explosion as

$$AB \rightarrow A^{m+} + B^{n+} + (m+n)e^-.$$

It is experimentally recognized that (i) Coulomb explosion takes place at a specific geometrical structure for diatomic and linear molecules and (ii) the specific geometrical structure is independent of any ionization stage during ionization steps. This is proved by a comparison of the measured kinetic energy of the fragment ions with the energy estimated by the Coulomb law. For diatomic molecules, if Coulomb explosion takes place at the equilibrium internuclear distance R_e, its energy E_e is given as $E_e = eQ_A eQ_B/R_e$, where $Q_A(Q_B)$ is the fragment A (B) ion charge. For example, for an iodine molecule I_2, the observed Coulomb energy is 70% of that expected from the Coulomb law at $R_e = 2.67\,\text{Å}$.[36] From the observed Coulomb energy, the Coulomb explosion takes place at a critical internuclear distance at $R_c = 3.8\,\text{Å}$.

The mechanism of Coulomb explosion of a diatomic molecule is qualitatively explained in terms of the classical model as shown in Fig. 10. Let first the molecule be ionized by a laser with polarization parallel to the molecular axis. Concerning the experimental evidence (i), consider the molecular ion dissociating, as the electric field amplitude of laser increases. The molecular ion consists of the outer electron and the other electrons and two nuclei. The outer electron is strongly influenced by laser, while the other electrons and nuclei are not. These particles act as ion cores. The double-well electron potential energy curve U is given as

$$U = -\frac{e^2 Q_A}{\left|x + \frac{R_c}{2}\right|} - \frac{e^2 Q_B}{\left|x - \frac{R_c}{2}\right|} - Fx,$$

where $Q_A(Q_B)$ is a charge of fragment A(B) and F is the laser electric field amplitude.

$U_I(U_0)$ in Fig. 10 denotes the inner (outer) potential. The energy level E_L of the outer electron in the double well is approximately expressed by taking averaging over the electronic energy at each site as

$$E_L = \frac{1}{2}\left(-E_A - \frac{e^2 Q_B}{R_c} - E_B - \frac{e^2 Q_A}{R_c}\right),$$

where $E_A(E_B)$ is the ionization potential of fragment ion A(B). This is lowered by the Coulomb potential of its neighboring ion, $\frac{e^2 Q_B}{R_c}$. In the above derivation, electric field effects on the relationship among E_L, U_I, and U_0 are omitted.

The key point for explaining the experimental evidence (i) is that the inner barrier height depends on the internuclear distance R. The barrier height of the inner potential U_I is lower than that of the outer potential barrier U_0 at R_e. In this case, the electron follows adiabatically the electric field of the laser pulse and goes back and forth between the two ion cores. The barrier height of U_I becomes higher, as R becomes longer. When U_I becomes equal to E_L at a critical bond length R_c, the dissociative ionization, i.e., Coulomb explosion, takes place.

Concerning the experimental evidence (ii), let us consider the molecule is ionized further, as the laser field strength increases. The next electron is in lower energy level. At the same time, the atomic potential barrier is also lower. The two effects combine

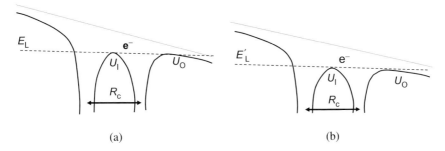

Fig. 10. Classical model of the Coulomb explosion of a diatomic molecule. Bold curves denote effective Coulomb potential energies (sum of electronic energy and molecule–laser field interaction energy indicated by a dotted line) of the diatomic ion. (a) The outer electron with energy E_L in the double-well potential at a critical internuclear distance R_c. (b) The next electron with $E'_L (<E_L)$ at the same R_c as that in (a) after the outer electron released. U_I and U_O are the inner and outer potential, respectively. Coulomb explosion occurs at the critical internuclear distance R_c where U_I and U_O are equal each other and also equal to the electron energy E_L.

to reproduce the same situation as the outer electron as shown in Fig. 10(b). In other words, for the next electron, the critical distance, R_c, is virtually unchanged. This behavior continues for further ionization processes. This qualitatively explains the experimental evidence (ii).

It should be noted that in the classical model, the effects of electronic dynamics are not taken into account at all. As shown in Chap. 5, electron–electron correlated motions play an important role in Coulomb explosions.

A quantitative explanation of Coulomb explosion-based quantum dynamical treatment and discussion of the mechanisms of Coulomb explosion of polyatomic molecules will be given in Chap. 5. Coulomb explosion has a possible application to imaging of molecules with sub-Angstrom spatial resolution within sub-femtoseconds.[37]

10. Alignment and Orientation

Since most molecules are anisotropic quantum systems, molecular alignment and orientation dependence in various phenomena are always a matter of central concern. Therefore, it is desirable to prepare a sample of aligned or oriented molecules, because important information about alignment and orientation dependence is generally lost in experiments with a sample of randomly oriented molecules. As shown in Fig. 11, "alignment" means a state where the molecular axes are parallel to each other without paying attention to the head-versus-tail directions. On the other hand, "orientation" means a state where the molecular axes are parallel to each other with their heads directed the same way.

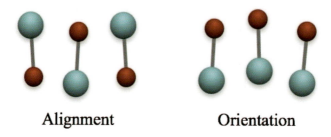

Alignment   Orientation

Fig. 11. Alignment and orientation of molecules.

Historically, stereodynamics in chemical reactions has been investigated with a sample of oriented molecules,[38,39] which can be prepared by two traditional techniques of brute-force orientation[40,41] and hexapole focusing.[42] In the brute-force orientation, a very high electrostatic field can be used to orient some specific molecules with large permanent dipole moments. An inhomogeneous electrostatic field formed by a hexapole focuser can be used to select and focus (near) symmetric top molecules in a specific rotational quantum state, which can be further oriented by an electrostatic field. Therefore, even with the above two traditional techniques, those molecules that can be oriented are quite limited and a vast number of nonpolar molecules cannot be aligned.

The modern version of controlling molecular alignment and orientation is based on the anisotropic polarizability interaction, *i.e.*, the interaction of an induced dipole moment in a molecule with an intense nonresonant pulsed laser field.[39,43,44] The spatial direction of a molecule is determined by a set of three Euler angles (Fig. 12).

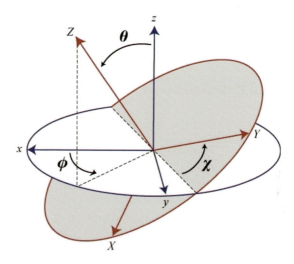

Fig. 12. The relative rotation between the body-fixed (X, Y, Z) and space-fixed (x, y, z) coordinate systems is specified by a set of three Euler angles (θ, ϕ, χ).

Controlling one of the three Euler angles is called 1D control. When all the three Euler angles are controlled, the technique is called 3D control, which is one of the remarkable advantages in the laser-based molecular manipulation techniques and cannot be achieved with the above traditional techniques.

In general, the laser-induced molecular alignment and orientation techniques are classified into the adiabatic and nonadiabatic regimes. When the duration of a laser pulse is much longer than the rotational period of a molecule, the molecular alignment and orientation are achieved in an intense nonresonant laser field. The molecular alignment and orientation are gradually (literally adiabatically) achieved in the rising part of a longer laser pulse and they evolve back to free rotational states after the pulse passes. 1D and 3D alignments have been achieved with intense nonrensonat linearly and elliptically polarized laser fields, respectively.[45,46] On the other hand, 1D and 3D orientations have been achieved with intense nonresonant linearly and elliptically polarized laser fields in a moderate electrostatic field.[47,48]

When molecules are subject to an intense, nonresonant, and ultrashort (typically femtosecond) laser pulse, rotational wavepackets are created by an impulsive Raman process. Field-free 1D molecular alignment is achieved, as quantum revivals of the rotational wavepackets created by an intense ultrashort linearly polarized laser field.[49] Field-free 3D molecular alignment has been demonstrated by using two time-delayed, orthogonally polarized, and femtosecond laser pulses.[50] On the other hand, laser-field-free 1D molecular orientation has recently been demonstrated with combined electrostatic and intense, rapidly turned-off, linearly polarized laser fields.[51] By employing an intense and rapidly turned-off laser field with elliptical polarization instead of linear polarization in a moderate electrostatic field, laser-field-free 3D alignment and orientation could be achieved just after the truncation of the laser pulse.[51,52]

The basic experimental techniques of molecular alignment and orientation are discussed in Chap. 2. The details about molecular alignment and orientation and recent representative studies with a sample of aligned or oriented molecules are discussed in Chap. 4, which forms one of the features of this monograph.

11. Molecular Chirality

Molecular chirality or molecular handedness is the characteristic of molecules having no inversion center or plane of symmetry.[53] Biomolecules in nature have this characteristic. Synthesis of chiral molecules called enantiomers has been the main subject in stereochemistry and biochemistry.[54] Enantiomers are produced by using catalytic reagents in laboratories and industrials.

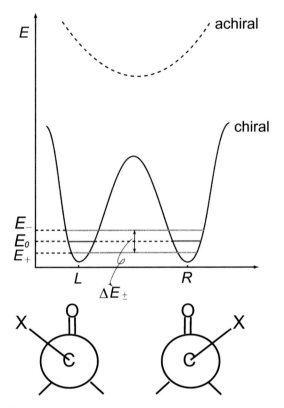

Fig. 13. A quantum mechanical treatment of chiral transformation proposed by Hund. A double-well potential model of transformation between L- and R-enantiomers is adopted by neglecting the parity-violating potential effects. Enantiomers in the ground state are mirror image with respect a mirror setting at the origin each other. C–H bond rotation around the center is the reaction coordinate. Energy splitting due to tunneling between the lowest states of the enantiomers is given by $\Delta E_\pm$.

Figure 13 shows a quantum mechanical treatment of molecular charity first reported by Hund in 1927.[55] Chiral transformation is viewed as tunneling in a double-well potential. Here, potential energies of pure enantiomers, L (left-handed) and R (right-handed) forms, are plotted as function of a reaction path of the chirality transformation. Rotation of the chemical bond C–X is taken as the reaction coordinate. The localized states of L- and R-enantiomers, χ_L and χ_R, are expressed in a linear combination of the eigenstates in the stationary state of the system, ψ_+ and ψ_-, as

$$\chi_L = \frac{1}{\sqrt{2}}(\psi_+ + \psi_-) \quad \text{and} \quad \chi_R = \frac{1}{\sqrt{2}}(\psi_+ - \psi_-).$$

If R-enantiomer is prepared at a specific time, then the R-enantiomer is transferred into L-enantiomer within time $t = \hbar/(2\Delta E_\pm)$ due to tunneling. Here, $\Delta E_\pm$ is the energy difference between the two eigenstates. As a result of the tunneling, L- and R-enantiomers are obtained with equal probability. This system is called racemic mixture. In the original treatment by Hund, the energy difference between two enantiomers was omitted. In actual systems, the potential energies of L- and R-enantiomers are not equal each other because of the parity-violating interaction term.[56] However, its magnitudes are about an order of 10^{-13} cm^{-1},[57] and so the contribution of the parity violating force can be safely omitted in ordinary time scale for chirality transformation.

In Chap. 8, a new scenario of chirality transformation by using lasers, quantum control of molecular chirality, will be described.

It should be noted that in Fig. 13 that the potential energy function of each enantiomer is asymmetric with respect to reflection of a mirror set at the bottom. This asymmetry induces unidirectional nuclear motions driven by a linearly polarized electric field of laser. An application of this idea to the control of molecular motors will be presented in Chap. 8. Molecular motors are one of the microscopic molecular functions. It is interesting to induce such a molecular function from nuclear dynamics.[58]

12. Molecular ADK Theory for Tunneling Ionization Rate

The ADK (Ammosov–Delone–Krainov) theory[59(a)] for the calculation of tunneling ionization rate of atoms is based on a simple model in which the atomic states at the tunneling point are approximated by those of a hydrogen-like atom. Molecules are characterized by a multi-centered body with vibrational and rotational degrees of freedoms, while atoms are characterized by a one-centered body. Tong *et al.* developed molecular tunneling ionization theory, called molecular ADK theory.[59(b)] In the molecular ADK theory, electronic wavefunctions in the asymptotic region are expressed in terms of the summation of spherical harmonics in a one-center expansion.

Let a molecule be aligned along the electric field of an intense laser. The molecular wavefunction at a large distance where tunneling occurs can be written as

$$\Psi^m(r) = \sum_l C_l F_l(r) Y_{lm}(\hat{r}). \tag{10}$$

Here, $F_l(r)$, atomic wavefunction of the valence electron in the asymptotic region in atomic units, is written as

$$F_l(r \to \infty) \approx r^{Z_c/\kappa - 1} \exp(-\kappa r), \tag{11}$$

where Z_c is the effective Coulomb charge, $\kappa = \sqrt{2I_p}$ with ionization potential I_p, and $Y_{lm}(\hat{r})$ is the spherical harmonics with angular quantum numbers l and m. The coefficients C_l are determined by fitting the asymptotic molecular wavefunction that is calculated by quantum chemistry procedures in the asymptotic form. The tunneling ionization rate for a diatomic molecule in a static field with strength F is given as

$$w_{\text{stat}}(F, 0) = \frac{B^2(m)}{2^{|m|}|m|!} \frac{1}{\kappa^{2Z_c/\kappa - 1}} \left(\frac{2\kappa^3}{F}\right)^{2Z_c/\kappa - |m| - 1}$$
$$\times \exp\left(-\frac{2\kappa^3}{3F}\right). \qquad (12)$$

Here, $B(m) = \sum_l C_l Q(l, m)$ with $Q(l, m) = (-1)^m \left(\frac{(2l+1)(l+|m|)!}{2(l-|m|)!}\right)^{\frac{1}{2}}$.

If a diatomic molecule is aligned not along the field direction but at an arbitrary angle $\boldsymbol{R}$ with respect to the field direction, the static field ionization rate is given as

$$w_{\text{stat}}(F, \boldsymbol{R}) = \sum_{m'} \frac{B^2(m')}{2^{|m'|}|m'|!} \frac{1}{\kappa^{2Z_c/\kappa - 1}} \left(\frac{2\kappa^3}{F}\right)^{2Z_c/\kappa - |m'| - 1}$$
$$\times \exp\left(-\frac{2\kappa^3}{3F}\right). \qquad (13)$$

Here, $B(m') = \sum_l C_l D^l_{m',m}(\boldsymbol{R}) Q(l, m')$, where $D^l_{m',m}(\boldsymbol{R})$ is the rotation matrix and $\boldsymbol{R}$ is the Euler angle between the diatomic molecular axis and the field direction.

The ionization rate in a low-frequency laser field is obtained after taking the average of the rates over an optical cycle as

$$w(F, \boldsymbol{R}) = \left(\frac{3F}{\pi\kappa^3}\right)^{\frac{1}{2}} w_{\text{stat}}(F, \boldsymbol{R}), \qquad (14)$$

where F denotes the peak strength of the laser field.

Figure 14 shows the H_2 ionization rates obtained by using Eq. (14). The ionization rates calculated by the molecular ADK theory are in agreement with those obtained from an *ab initio* calculation. Here, the correlation effects of the two electrons are taken into account.

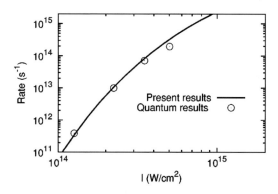

Fig. 14. Ionization rate of H_2 in an intense laser field at the equilibrium distance. The circles are from the *ab initio* molecular orbital calculations. (Reprinted with permission from Tong et al.[59(b)] Copyright (2002) by the American Physical Society).

A generalization and extension of the Keldysh theory to molecules have been reported by several groups.[61] In Keldysh's original work,[16(a)] an approximate expression for the tunneling ionization rate of one-electron atoms was derived within the first-order time-dependent perturbation theory. Time integration was performed by using the saddle point integration method. An analytical expression for the tunneling ionization rate of one-electron atoms has been derived by using a residue integration method in the complex plane by Mishima et al.[61(a)] Numerical calculations showed that the calculated ionization rate for hydrogen atoms is greater than that in Keldysh's original results, which is accordance with the experimental results. They have extended the derivation to ionization rates of randomly oriented diatomic and polyatomic molecules.[61(a)-(f)]

13. Summary

In this chapter, several concepts and phenomena associated with electronic and nuclear dynamics of molecules are explained with simplified models. These are starting points in quantum dynamics of electrons and nuclei of molecules. Detailed treatments in electronic and nuclear dynamics of molecules will be described in details in the latter chapters.

References

1. D. J. Tannor, *Introduction to Quantum Mechanics: A Time-Dependent Perspective* (University Science Books, Sausalito California, 2007).
2. J. C. Polanyi and A. H. Zewail, *Acc. Chem. Res.* **28**, 119 (1995).
3. (a) A. Stolow, *Phil. Trans. R. Soc. Lond. A* **356**, 345 (1998); (b) K. Ohmori *Annu. Rev. Phys. Chem.* **60**, 487 (2009).
4. (a) J. S. Baskin, P. M. Felker and A. H. Zewail, *J. Chem. Phys.* **84**, 4708 (1986); (b) J. A. Yeazell and C. R. Stround Jr., *Phys. Rev. A* **43**, 5153 (1991); (c) T. Fujiwara, Y. Fujimura and O. Kajimoto, *Chem. Phys. Lett.* **261**, 201 (1996); (d) T. Seideman, *Phys. Rev. Lett.* **83**, 4971 (1999); (e) M. J. J. Vrakking, D. M. Villeneuve and A. Stolow, *Phys. Rev. A* **54**, R37 (1996); (f) Ch. Warmuth, A. Tortschanoff, F. Milota, M. Leibscher, M. Shapiro, Y. Prior, I. Sh, Averbukh, W. Schleich, W. Jakubetz and H. F. Kauffmann, *J. Chem. Phys.* **114**, 9901 (2001).
5. T. S. Rose, M. J. Rosker and A. H. Zewail, *J. Chem. Phys.* **88**, 6672 (1988).
6. (a) P. M. Felker and A. H. Zewail, *J. Chem. Phys.* **82**, 2961 (1985); (b) K. Ohmori, H. Katsuki, H. Chiba, M. Honda, Y. Hagihara, K. Fujiwara, Y. Sato and K. Ueda, *Phys. Rev. Lett.* **96**, 093002 (2006).
7. A. D. Bandrauk and H. Kono, in S. H. Lin, A. A. Villaeys and Y. Fujimura (Eds.), *Adv. Multiphoton and Spectroscopy*, Vol. 15 (World Scientific, Singapore, 2003), p. 149.
8. S. Teufel, *Adiabatic Perturbation Theory in Quantum Dynamics* (Springer-Verlag, Berlin, 2003).

9. S. H. Lin, *J. Chem. Phys.* **44**, 3759 (1966).
10. (a) B. Kohler, J. L. Krause, F. Raksi, K. R. Wilsion, V. V. Yakovlev, R. M. Whitnell and Y. Yan, *Acc. Chem. Res.* **28**, 133 (1995); (b) S. A. Rice and M. Zhao, *Optical Control of Molecular Dynamics* (John Wiley & Sons Inc., New York, 2000).
11. (a) R. J. Gordon and Y. Fujimura, *Coherent Control of Chemical Reactions* (Academic Press, San Diego, 2002), Encyclopedia of Physical Science and Technology Vol. 3, p. 207; (b) A. D. Bandrauk, Y. Fujimura and R. J. Gordon, Eds., *Laser Control and Manipulation of Molecules*, No. 821 in CS Symposium Series (American Chemical Society, Washington D.C., 2002).
12. R. Uberna, Z. Amitay, R. A. Loomis and S. R. Leone, *Faraday Discuss.* **113**, 385 (1999).
13. A. Assion, T. Baumert, M. Bergt, T. Brixner, B. Kiefer, V. Seyfried, M. Strehle and G. Gerber, *Science* **282**, 919 (1998).
14. S. Shi and H. Rabitz, *J. Chem. Phys.* **92**, 364 (1990).
15. (a) N. B. Delone and V. P. Krainov, *J. Opt. Soc. Am. B* **8**, 1207 (1991); (b) N. B. Delone and V. P. Krainov, *Multiphoton Processes in Atoms* (Springer-Verlag, Berlin, 1994).
16. (a) L. Keldysh, *Zh. Eksp. Teor. Fiz.* [*Sov. Phys. JETP* **20**, 1307 (1965)]; (b) M. Gavrila, *Atoms in Intense Laser Fields* (Academic Press, New York, 1992); (c) S. L. Chin, in S. H. Lin, A. A. Villaeys and Y. Fujimura (Eds.), *Adv. Multiphoton and Spectroscopy*, Vol. 16 (World Scientific, Singapore, 2004), p. 249.
17. (a) M. Protopapas, C. H. Keitel and P. L. Knight, *Rep. Prog. Phys.* **60**, 389 (1997); (b) P. Sauères, A. L'Huillier, P. Antoine and M. Lewenstein, *Adv. At. Mol. Opt. Phys.* **41**, 83 (1999).
18. Y. Nabekawa and K. Midorikawa, in S. H. Lin, A. A. Villaeys and Y. Fujimura (Eds), *Adv. Multiphoton and Spectroscopy*, Vol. 18 (World Scientific, Singapore, 2008), p. 1.
19. (a) A. McPherson, G. Gibson, H. Jara, U. Johann, T. S. Luk, I. A. McIntyre, K. Boyer and C. K. Rhodes, *J. Opt. Soc. Am. B* **4**, 595 (1987); (b) A. L'Huillier, K. J. Schafer, and K. C. Kulander, *J. Phys. B At. Mol. Opt. Phys.* **24**, 3315 (1991).
20. K. Miyazaki and H. Sakai, *J. Phys. B At. Mol. Opt. Phys.* **25**, L83 (1992).
21. (a) Y. Liang, S. J. Augst and S. L. Chin, *Opt. Photon News Phys.* **5**, 60 (1994); (b) Y. Liang, S. J. Augst, S. L. Chin, Y. Beaudoin and M. Chaker, *J. Phys. B At. Mol. Opt. Phys.* **27**, 5119 (1994).
22. J. L. Krause, K. J. Schafer and K. C. Kulander, *Chem. Phys. Lett.* **178**, 573 (1991).
23. T. Zuo, S. Chelkowski and A. D. Bandrauk, *Phys. Rev. A* **48**, 3837 (1993).
24. C. Lyngå, A. L'Huillier and C.-G. Wahlström, *J. Phys. B At. Mol. Opt. Phys.* **29**, 3293 (1996).
25. (a) R. Velotta, N. Hay, M. B. Mason, M. Castillejo and J. P. Marangos, *Phys. Rev. Lett.* **87**, 183901(2001); (b) J. Itatani, D. Zeidler, J. Levesque, M. Spanner, D. M. Villeneuve and P. B. Corkum, *Phys. Rev. Lett.* **94**, 123902 (2005); (c) M. Lein and C. Chirila, in S. H. Lin, A. A. Villaeys and Y. Fujimura (Eds), *Adv. Multiphoton and Spectroscopy*, Vol. 18 (World Scientific, Singapore, 2008), p. 69.
26. P. B. Corkum, *Phys. Rev. Lett.* **71**, 1994 (1993).
27. A. D. Bandrauk and H. Yu, *Phys. Rev. A* **59**, 539 (1999); (b) A. D. Bandrauk, S. Barmaki, S. Chelkowski and G. L. Kamta, in K. Yamanouchi, S. L. Chin, P. Agostini and G. Ferrante (Eds), *Prog. in Ultrafast Intense Laser Science III*, (Springer-Verlag, Berlin, 2008), p. 171.

28. (a) S. H. Lin, Y. Fujimura, H. J. Neuser and E. W. Schlag, *Multiphoton Spectroscopy of Molecules* (Academic Press, Orland, 1984); (b) H. Niikura and P. B. Corkum, *Adv. At. Mol. Opt. Phys.* **54**, 511 (2007).
29. P. Agostini, F. Fabre, G. Mainfray, G. Petite and N. K. Rahman, *Phys. Rev. Lett.* **42**, 1127 (1979).
30. C. Cornaggia, J. Molellec and D. Normand, *J. Phys. B At. Mol. Opt. Phys.* **18**, L501 (1985).
31. C. Cornaggia, D. Normand, J. Molellec, G. Mainfray and C. Manus, *Phys. Rev. A* **34**, 207 (1986).
32. A. D. Bandrauk, Ed., *Molecules in Laser Fields* (Marcel Dekker, New York, 1994).
33. (a) P. H. Bucksbaum, A. Zavriyev, H. G. Muller and D. W. Schumacher, *Phys. Rev. Lett.* **64**, 1883 (1990); (b) A. Giusti-Suzor, X. He, O. Atabek and F. H. Mies, *Phys. Rev. Lett.* **64**, 515 (1990); (c) M. Sugawara, M. Kato and Y. Fujimura, *Bull. Chem. Soc. Jpn.* **66**, 3253 (1993); (d) K. Sugimori, T. Ito, Y. Takata, K. Ichitani, H. Nagao and K. Nishikawa, *J. Phys. Chem. A* **111**, 9417 (2007).
34. D. Normand, C. Cornaggia and J. Molellec, *J. Phys. B At. Mol. Phys.* **19**, 2881 (1986).
35. (a) L. J. Fransinsky, K. Codling, P. Hatherly, J. Barr, I. N. Ross and W. T. Toner, *Phys. Rev. Lett.* **58**, 2424 (1987); (b) C. Cornaggia, D. Normand and J. Morellec, *J. Phys. B At. Mol. Opt. Phys.* **25**, L415 (1992); (c) A. Hishikawa, A. Iwamae, K. Hoshina, M. Kono and K. Yamanouchi, *Chem. Phys. Lett.* **282**, 283 (1998); (d) M. J. Dewitt and R. J. Levis, *Phys. Rev. Lett.* **81**, 5101 (1998); (e) N. Nakashima, T. Yatsuhasi, M. Murakami, R. Mizoguchi and Y. Shimada in S. H. Lin, A. A. Villaeys and Y. Fujimura (Eds.), *Adv. Multiphoton and Spectroscopy*, Vol. 17 (World Scientific, Singapore, 2006), p. 179.
36. J. H. Postumus, L. J. Fransinski, A. J. Giles and K. Codling, *J. Phys. B At. Mol. Opt. Phys.* **28**, L349 (1995).
37. G. L. Kamta and A. D. Badrauk, *Phys. Rev. A* **74**, 033415 (2006).
38. Special issue on Stereodynamics of Chemical Reactions H. Loesch (Organizer) [*J. Phys. Chem. A* **101**, 7461 (1997)].
39. D. Herschbach, *Eur. Phys. J. D* **38**, 3 (2006).
40. B. Friedrich and D. R. Herschbach, *Nature* (London) **353**, 412 (1991).
41. H. J. Loesch and A. Remscheid, *J. Phys. Chem.* **95**, 8194 (1991).
42. V. A. Cho and R. B. Bernstein, *J. Phys. Chem.* **95**, 8129 (1991).
43. H. Stapelfeldt and T. Seideman, *Rev. Mod. Phys.* **75**, 543 (2003).
44. T. Seideman and E. Hamilton, *Adv. At. Mol. Opt. Phys.* **52**, 289 (2005).
45. H. Sakai, C. P. Safvan, J. J. Larsen, K. M. Hilligsφe, K. Hald and H. Stapelfeldt, *J. Chem. Phys.* **110**, 10235 (1999).
46. J. J. Larsen, K. Hald, N. Bjerre, H. Stapelfeldt and T. Seideman, *Phys. Rev. Lett.* **85**, 2470 (2000).
47. H. Sakai, S. Minemoto, H. Nanjo, H. Tanji and T. Suzuki, *Phys. Rev. Lett.* **90**, 083001 (2003).
48. H. Tanji, S. Minemoto and H. Sakai, *Phys. Rev. A* **72**, 063401 (2005).
49. F. Rosca-Pruna and M. J. J. Vrakking, *Phys. Rev. Lett.* **87**, 153902 (2001).
50. K. F. Lee, D. M. Villeneuve, P. B. Corkum, A. Stolow and J. G. Underwood, *Phys. Rev. Lett.* **97**, 173001 (2006).
51. A. Goban, S. Minemoto and H. Sakai, *Phys. Rev. Lett.* **101**, 013001 (2008).
52. Y. Sugawara, A. Goban, S. Minemoto and H. Sakai, *Phys. Rev. A* **77**, 031403(R) (2008).

53. (a) M. Avalos, R. Babiano, P. Cintas, J. Jiménez, J. C. Palacios and L. D. Barron, *Chem. Rev.* **98**, 2391 (1998); (b) S. Aono, *J. Phys. Soc. Jpn.* **73**, 2712 (2004).
54. R. Noyori, *Asymmetric Catalysis in Organic Synthesis* (Wiley, New York, 1994).
55. (a) F. Hund, *Z. Phys.* **43**, 788 (1927); (b) *ibid.* **43**, 805 (1927).
56. T. D. Lee and C. N. Yang, *Phys. Rev.* **104**, 254 (1956).
57. M. Quack, *Angew. Chem. Int. Ed.* **41**, 4618 (2002).
58. M. Yamaki, S. Nakayama, K. Hoki, H. Kono and Y. Fujimura, *Phys. Chem. Chem. Phys.* **11**, 1662 (2009).
59. (a) M. V. Ammosov, N. B. Delone and V. P. Krainiv, *Zh. Exsp. Teor. Fiz.* **91**, 2008 (1986) [*Sov. Phys. JETP* **64**, 1191 (1987)]; (b) X. M. Tong, Z. X. Zhao and C. D. Lin, *Phys. Rev. A* **66**, 033402 (2002).
60. A. Saenz, *Phys. Rev. A* **61**, 051402(R) (2000).
61. (a) K. Mishima, M. Hayashi, J. Yi, S. H. Lin, H. L. Selzle and E. W. Schlag, *Phys. Rev. A* **66**, 033401 (2002); (b) K. Mishima, K. Nagaya, M. Hayashi and S. H. Lin, *Phys. Rev. A* **70**, 063414 (2004); (c) K. Mishima, K. Nagaya, M. Hayashi and S. H. Lin, *J. Chem. Phys.* **122**, 104312 (2005); (d) K. Mishima, K. Nagaya, M. Hayashi and S. H. Lin, *J. Chem. Phys.* **122**, 024104 (2005); (e) K. Mishima, M. Hayashi and S. H. Lin, *Phys. Rev. A* **71**, 053411 (2005). (f) H. Mineo, S. D. Chao, K. Mishima, K. Nagaya, M. Hayashi and S. H. Lin, *Phys. Rev. A* **75**, 027402 (2007); (g) K. Mishima, K. Nagaya, M. Hayashi, S. H. Lin and E. W. Schlag, in S. H. Lin, A. A. Villaeys and Y. Fujimura (Eds.), *Adv. Multiphoton and Spectroscopy*, Vol. 17 (World Scientific, Singapore, 2006), p. 29; (h) S. D. Chao, *Phys. Rev. A* **72**, 053414 (2005).

Chapter 2

Experimental Setups and Methods

In this chapter, some basic experimental setups and methods related to this monograph are described, so that the readers can understand the experimental discussions in the following chapters. The topics include laser-induced Coulomb explosion imaging, high-order harmonic generation, and molecular alignment and orientation.

1. Laser-Induced Coulomb Explosion Imaging

When molecules are irradiated with an intense femtosecond laser pulse, they are generally multiply ionized and rapidly dissociated into fragment ions due to Coulombic repulsion, which is called Coulomb explosion. Since the process is so rapid, the directions of the emitted fragment ions are thought to reflect the structure of the sample molecules just before the Coulomb explosion. In the pioneering studies of Coulomb explosion imaging, Coulomb explosion was induced by the collision between a fast molecular beam and a thin foil and was used to extract structural information of molecules.[1,2] The modern version of the technique is based on the laser-induced Coulomb explosion, which allows us to employ the pump-probe technique and thereby to follow the ultrafast dynamics of structural changes of molecules.[3]

A velocity map imaging spectrometer including a set of three aperture electrodes is used to observe fragment ions produced by Coulomb explosion.[4] A velocity map condition can be met by carefully adjusting applied voltages to the repeller and extractor. Thereby, the ions with the same initial velocity vector can be mapped onto the same position on the detector plane because of the electrostatic lens effect. Some problems in conventional spectrometers with grid electrodes such as transmission reduction, trajectory deflections, and blurring of fragment ions can be avoided by the new spectrometers with aperture electrodes. Photoelectrons can also be observed with the same spectrometer by changing the applied voltages to the opposite polarity.[5]

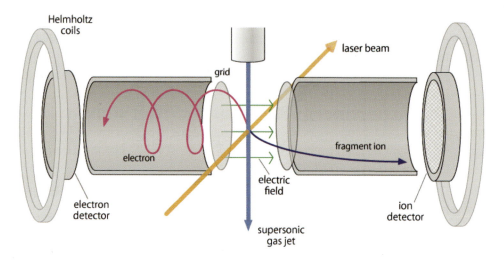

Fig. 1. A schematic diagram of a reaction microscope.

A further advanced device for imaging studies is a reaction microscope where simultaneous recoil-ion and electron momentum imaging is performed (Fig. 1).[6] In a reaction microscope, so-called three-dimensional (3D) detectors are placed on both sides of the device to determine the initial momentum vectors of ions and electrons unambiguously based on their position and time-of-flight information. In order to achieve an almost 4π collection efficiency for electrons with high kinetic energies of more than 100 eV, Helmholtz coils are usually used to generate a homogeneous magnetic field and make electron trajectories spiral. Although the name of COLTRIMS originally stands for COLd Target Recoil-Ion Momentum Spectroscopy, which originates from the combination of cold targets of atoms and molecules prepared by a supersonic gas-jet valve and recoil-ion momentum spectroscopy (RIMS), it is sometimes used as a synonym for simultaneous recoil-ion and electron momentum spectroscopy. Therefore, a COLTRIMS apparatus sometimes means a reaction microscope. Although a reaction microscope is indeed a powerful tool for molecular imaging studies, a constraint is the limited number of processable charged particles in the device at a time.

2. High-Order Harmonic Generation

When atoms and molecules are exposed to an intense ultrashort laser pulse, high-order (usually odd-order because of inversion symmetry of a nonlinear medium) harmonics are generated. High-order harmonic generation (HHG) has been a subject of intensive and extensive studies for more than two decades by reason of its potential as a coherent ultrashort radiation source in the extreme ultraviolet (XUV) and soft x-ray regions.[7] Recently, researchers are especially interested in attosecond pulse generation[8,9] and molecular imaging based on HHG.[10]

The basic physics of HHG is well understood by the three-step model.[11] First, a part of the bound-state electron wave packet tunnels through the potential barrier modified by the intense laser field and emerges in the continuum (step 1). The freed electron wave packet is then driven by the laser field and after the field reverses its direction, it has a probability of returning to the parent ion (step 2). The high-energy photon is emitted because of the coherent oscillation between the returning electron wave packet and the remaining bound-state electron wave packet (step 3).

When aligned molecules are used as a nonlinear medium, a sort of pump-probe experiment is performed. A typical experimental setup is shown in Fig. 2[12] and the outline of an HHG experiment from aligned molecules is as follows: an output from a Ti:sapphire-based chirped pulse amplification system is split into two pulses with a Michelson-type interferometer. The first pulse is used as a pump to create rotational wave packets and induce nonadiabatic molecular alignment. The second pulse is delayed by a translation stage and is used as a probe to generate high-order harmonics. The two pulses are collinearly focused with the same lens into a supersonic molecular beam in the vacuum chamber. The generated harmonics are spectrally dispersed by a grazing incidence vacuum monochromator and detected by an x-ray charge-coupled device camera or an imaging detector consisting of a microchannel plate backed by a phosphor screen. In general, both pump and probe pulses are blocked before the monochromator by an appropriate XUV filter. The observed harmonic spectrum is corrected by taking account of the XUV filter transmittance, the grating reflection efficiency, and the quantum efficiency of the detector. In addition to the observation of harmonics, it is quite useful to detect ions produced through multiphoton ionization by an ion collector prepared downstream of the gas jet (see Fig. 2), so that one can disentangle the contributions from the ionization (step 1) and recombination (step 3) processes.[13] The propagation effect

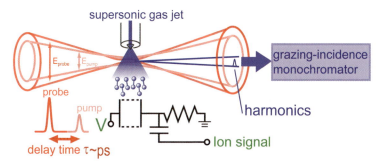

Fig. 2. An experimental setup for high-order harmonic generation from aligned molecules. (Reprinted with permission from Kanai.[12])

may influence harmonic generation and can be controlled by adjusting the focusing geometry of the probe pulse relative to a supersonic molecular beam.[14]

3. Molecular Orientation and Its Observation

In this section, the techniques to achieve and observe 1D and 3D molecular orientations are described based on Refs. 15–17. The technique for 3D orientation includes that for molecular alignment.[18,19] Although the techniques described here are those in the adiabatic regime, the temporal evolutions of molecular alignment and orientation in the nonadiabatic regime can be followed by applying the pump-probe technique. The theoretical background for molecular alignment and orientation can be found in Chap. 4.

3.1 1D molecular orientation

Figure 3 shows a schematic diagram of the experimental setup. A pulsed supersonic beam of OCS molecules (permanent dipole moment $\mu = 0.71$ D)[20] is supplied by expanding OCS molecules diluted (5%) with argon or helium carrier gas through a 0.25-mm-diameter nozzle. The molecular beam is crossed at 90° by the focused laser pulses. To achieve molecular orientation, an extraction field of the time-of-flight (TOF) spectrometer is used for the interaction between the permanent dipole and the electrostatic field. The molecular orientation can be greatly enhanced by the anisotropic polarizability interaction between the induced dipole moment and an intense nonresonant linearly polarized laser pulse,[21,22] for which the fundamental

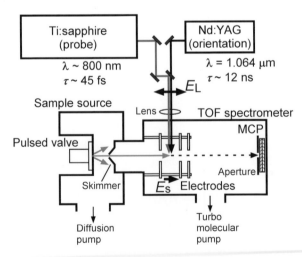

Fig. 3. A schematic diagram of the experimental setup for 1D molecular orientation. (Reprinted with permission from Minemoto et al.[16] Copyright (2003) by the American Institute of Physics.)

field (wavelength $\lambda = 1064$ nm) from an injection-seeded Nd:YAG laser is used. The polarization of the YAG pulse is parallel to the electrostatic field, *i.e.*, the TOF axis. The duration of the YAG pulse is ~ 12 ns (full width at half maximum) and long enough to ensure that the experiments are performed in the adiabatic regime.

An intense femtosecond laser pulse is used as a probe pulse to multiply ionize the OCS molecules at the peak of the YAG pulse to evaluate the degree of orientation. The femtosecond probe pulses are delivered by a Ti:sapphire-based chirped pulse amplification system with the center wavelength of ~ 800 nm and the duration of ~ 45 fs. The both pulses are spatially overlapped using a dichroic mirror and focused by a 30-cm-focal-length lens into the interaction region of the TOF spectrometer. It is important to ensure that only those molecules that have been exposed to the YAG pulse are probed. Therefore, the focal spot size of the probe pulse is carefully adjusted to be smaller than that of the YAG pulse. The fragment ions produced by the probe pulse are detected by a microchannel plate (MCP) positioned on-axis with the TOF axis.

When the polarization of the probe pulse is parallel to the TOF axis, one usually observes a pair of peaks consisting of so-called "forward" and "backward" fragments whose initial velocities are directed toward and away from the MCP detector, respectively. This is because of enhanced ionization by which molecules initially aligned along the polarization are much more easily to be multiply ionized. Figure 4 shows typical TOF spectra observed (a) with and (b) without YAG pulses. Here, the

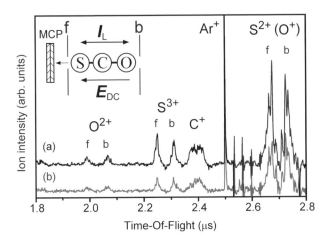

Fig. 4. Typical TOF spectra of OCS molecules (a) with and (b) without YAG pulses. The labels f and b denote fragment ions coming from the forward and backward initial emission directions, respectively. The inset shows the directions of the electrostatic field, the linearly polarized laser field, and an oriented OCS molecule. (Reprinted with permission from Sakai *et al.*[15] Copyright (2003) by the American Physical Society.)

attention is directed to the S^{3+} ion signals. When the YAG pulses are not present (Fig. 4(b)), the forward and backward signals look almost symmetric, indicating that the molecules are randomly oriented. When the YAG pulses are applied (Fig. 4(a)), the signals look asymmetric. This observation can be interpreted as the result that more than half the OCS molecules are oriented with their S atoms directed toward the MCP detector. In fact, the observation in Fig. 4(a) is consistent with the theoretical expectation because the permanent dipole moment of an OCS molecule is directed toward the S atom and the molecules tend to be oriented in the way shown in the inset of Fig. 4. The enhancement of the backward signal is the result of the combined effect of molecular alignment and enhanced ionization, that is, some molecules are not oriented but just aligned with the application of the YAG pulses. The above interpretation can be further confirmed by observing the signal of a counterpart fragment, *e.g.*, CO^+. The backward signal becomes larger than the forward signal when the YAG pulses are applied.

3.2 *3D molecular orientation*

The interaction between a molecule and the combination of an electrostatic field and an elliptically polarized laser field creates 3D potential wells with the asymmetry along the electrostatic field, which can be used to achieve 3D molecular orientation. 3D molecular orientation can be confirmed by complementary observations composed of 2D ion imaging and TOF mass spectrometry, which are schematically shown in Fig. 5. 3D alignment can be confirmed by observing 2D ion

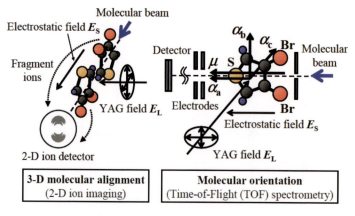

Fig. 5. Schematic diagrams of the complementary observations utilized for confirming 3D orientation. 3D alignment is confirmed with the 2D ion imaging technique on the left and 1D orientation is verified by TOF mass spectrometry on the right. μ and α represent the permanent dipole moment and the polarizability of a 3,4-dibromothiophene molecule, respectively. (Reprinted with permission from Tanji *et al.*[17] Copyright (2005) by the American Physical Society.)

images. Orientation can be confirmed by observing TOF spectra as in 1D molecular orientation described in Sec. 3.1. These two ways of observations provide us with firm evidence of 3D molecular orientation.

In order to achieve 3D orientation, the polarization of the YAG pulse is set to be elliptical. Ellipticity is defined as $\varepsilon = E_{\min}/E_{\maj}$, where $E_{\maj}$ and $E_{\min}$ are the major and minor axis components of the laser electric field, respectively. The ellipticity is adjusted by rotating a half-wave plate placed before a fixed quarter-wave plate. Thereby, one can utilize arbitrary ellipticity with the two axes of the ellipse fixed in the laboratory frame.

As a sample molecule, 3,4-dibromothiophene (DBT) molecules (illustrated in Fig. 5) are employed. A DBT molecule has a nonzero permanent dipole moment ($\mu = 1.59\,\text{D}$)[23] and three different polarizability components, which are required for the demonstration of 3D orientation. The molecular beam is supplied by expanding DBT molecules seeded in helium carrier gas into a vacuum chamber through a 0.5-mm-diameter nozzle of the pulsed valve. The applied voltages of the TOF spectrometer are adjusted for velocity mapping condition, so that the fragment ions with the same initial velocity arrive at the same position on the detector plane. The rest of the experimental details are basically the same as those in 1D orientation.

3D alignment achieved with an elliptically polarized laser field can be confirmed by observing ion images. The attention is directed to Br^+ and S^+ ion signals for confirming 3D alignment. Figure 6 shows typical images of Br^+ and S^+ ions observed by using various ellipticities of the YAG pulse. Here, the probe pulse is circularly polarized in order to avoid any influence of enhanced ionization on ion images.

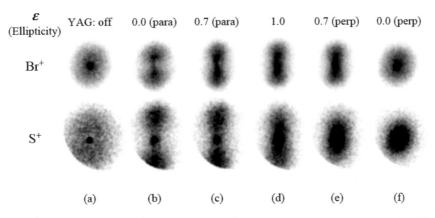

Fig. 6. Typical ion images of Br^+ (upper row) and S^+ (lower row) fragments observed with different ellipticities of the YAG pulses. The labels (para) and (perp) mean the major axis of the polarization ellipse parallel and perpendicular to the detector plane, respectively. (Reprinted with permission from Tanji et al.[17] Copyright (2005) by the American Physical Society.)

Figure 6(a) shows ion images observed without the YAG pulses. They look almost center-symmetric, indicating that the molecules are randomly oriented. When the YAG pulses are applied, distinct anisotropies are observed in the ion images (Figs. 6(b)–6(e)). When the YAG pulses are linearly polarized with the polarization parallel to the detector plane (Fig. 6(b)), both Br^+ and S^+ ions gather along the polarization and divide into top and bottom parts. Considering the structure of a DBT molecule illustrated in Fig. 5, the change observed in Fig. 6(b) can be interpreted as the evidence that the major axis of the molecule defined as the axis with the largest polarizability component is confined along the polarization, i.e., 1D alignment is achieved. Figure 6(f) observed with the polarization perpendicular to the detector plane serves to reinforce the interpretation of Fig. 6(b). The isotropic distributions of both Br^+ and S^+ ions show that the major axis of the molecule is now perpendicular to the detector plane.

Figure 6(d) observed with circular polarization shows that both ions are distributed along the vertical direction and the division into top and bottom parts has disappeared. Based on the theoretical predictions, this observation can be interpreted as the result that the molecular plane comes to be confined in the plane of circular polarization but the molecule is allowed to rotate in the polarization plane.

Figures 6(c) and 6(e) are observed with elliptical polarization ($\varepsilon = 0.7$) with the major axis of the ellipse parallel and perpendicular, respectively, to the detector plane. The images of S^+ ions shown in Figs. 6(c) and 6(e) indicate that the major axis of the molecule is confined parallel to that of elliptical polarization. In contrast, the images of Br^+ ions are distributed along the vertical direction. When Fig. 6(e) is compared to Fig. 6(f), it is noticed that the effect of elliptical polarization is distinct in squeezing the image of Br^+ ions in the horizontal direction and extending in the vertical direction. A series of images shown in Fig. 6 forms firm evidence that, under the application of an elliptically polarized laser field, the molecules are confined in the polarization plane with their major axes parallel to the major axis of the polarization ellipse, i.e., 3D alignment is achieved.

The next task is to confirm the "head-versus-tail" order of the molecules when an electrostatic field is also applied. For this purpose, the major axis of elliptical polarization is set to be perpendicular to the detector plane, i.e., parallel to the TOF axis, so that an extraction field of the TOF spectrometer can be used to achieve molecular orientation. By examining the S^+ ion signals as described in Sec. 3.1, it is confirmed that molecular orientation is achieved.

Thus, the complementary observations using 2D ion imaging and TOF mass spectrometry provide us with firm evidence that 3D molecular orientation can be achieved with combined electrostatic and elliptically polarized laser fields. An alternative approach for the confirmation of 3D molecular orientation is to observe asymmetry in the fragment yields of ion images by making an angle β between an electrostatic field and the major axis of elliptical polarization though only the cos β times the major axis component of elliptical polarization can contribute to molecular orientation.[24]

References

1. Z. Vager, R. Naaman and E. P. Kanter, *Science* **244**, 426 (1989).
2. Z. Vager, T. Graber, E. P. Kanter and D. Zajfman, *Phys. Rev. Lett.* **70**, 3549 (1995).
3. H. Stapelfeldt, E. Constant, H. Sakai and P. B. Corkum, *Phys. Rev. A* **58**, 426 (1998).
4. A. T. J. B. Eppink and D. H. Parker, *Rev. Sci. Instrum.* **68**, 3477 (1997).
5. T. Suzuki, *Annu. Rev. Phys. Chem.* **57**, 555 (2006).
6. J. Ullrich, R. Moshammer, A. Dorn, R. Dörner, L. Ph. H. Schmidt and H. Schmidt-Böcking, *Rep. Prog. Phys.* **66**, 1463 (2003).
7. J. G. Eden, *Prog. Quantum Electron.* **28**, 197 (2004).
8. P. Agostini and L. F. DiMauro, *Rep. Prog. Phys.* **67**, 813 (2004).
9. F. Krausz and M. Ivanov, *Rev. Mod. Phys.* **81**, 163 (2009).
10. M. Lein, *J. Phys. B* **40**, R135 (2007).
11. P. B. Corkum, *Phys. Rev. Lett.* **71**, 1994 (1993).
12. T. Kanai, Doctoral thesis, The University of Tokyo (2006).
13. T. Kanai, S. Minemoto and H. Sakai, *Nature* (London) **435**, 470 (2005).
14. P. Balcou, P. Salières, A. L'Huillier and M. Lewenstein, *Phys. Rev. A* **55**, 3204 (1997).
15. H. Sakai, S. Minemoto, H. Nanjo, H. Tanji and T. Suzuki, *Phys. Rev. Lett.* **90**, 083001 (2003).
16. S. Minemoto, H. Nanjo, H. Tanji, T. Suzuki and H. Sakai, *J. Chem. Phys.* **118**, 4052 (2003).
17. H. Tanji, S. Minemoto and H. Sakai, *Phys. Rev. A* **72**, 063401 (2005).
18. H. Stapelfeldt and T. Seideman, *Rev. Mod. Phys.* **75**, 543 (2003).
19. T. Seideman and E. Hamilton, *Adv. At. Mol. Opt. Phys.* **52**, 289 (2006).
20. A. A. Radzig and B. M. Smirnov, *Reference Data on Atoms, Molecules, and Ions* (Springer-Verlag, Berlin, Heidelberg, 1985).
21. B. Friedrich and D. Herschbach, *J. Chem. Phys.* **111**, 6157 (1999).
22. B. Friedrich and D. Herschbach, *J. Phys. Chem. A* **103**, 10280 (1999).
23. T. Shimozawa, *Bull. Chem. Soc. Jpn.* **38**, 1046 (1965).
24. A. Gijsbertsen, W. Siu, M. F. Kling, P. Johnsson, P. Jansen, S. Stolte and M. J. J. Vrakking, *Phys. Rev. Lett.* **99**, 213003 (2007).

Chapter 3

Theoretical Treatments of Wavepackets

In this chapter, theoretical treatments of electronic and nuclear wavepackets are briefly reviewed. First, the generation of wavepacket by pulse excitation and the propagation are described using a simple model based on the perturbative treatment. Second, numerical methods for propagation of wavepacket, which are valid for intense field cases as well as weak fields, are presented. Third, an incorporation of wavepacket propagation method into the scattering matrix method is presented for taking into account the effects of intramolecular electronic dynamics before ionization.

1. Generation of Wavepacket and Its Propagation

Wavepacket is a fundamental concept of ultrafast nuclear and/or electronic dynamics in molecules.[1] In this section, we examine how wavepacket is generated on an excited state potential energy surface by a laser pulse excitation and how they propagate. We use a simplified model in which a nuclear wavepacket is generated on a one-dimensional (1D) potential under a weak laser pulse condition. Molecular–laser field interactions are treated in the semiclassical way[2] *i.e.*, wavepacket is described quantum mechanically, while laser fields are described classically.

The total Hamiltonian is given in the semiclassical treatment as

$$H(t) = H_0 + V(t), \tag{1}$$

where H_0 denotes the molecular Hamiltonian and $V(t)$ the molecule–laser field interaction. The molecular Hamiltonian is expressed as

$$H_0 = T_R + T_r + U(r, R). \tag{2}$$

Here, r and R denote the coordinates of electrons and nuclei, respectively. T_R is the nuclear kinetic operator, T_r is the electronic kinetic operator, and $U(r, R)$ is the electrostatic interactions, which include the electron–nucleus, electron–electron, and nucleus–nucleus interactions.

The molecule–laser field interaction is given in the semiclassical treatment as

$$V(t) = -\boldsymbol{\mu} \cdot \boldsymbol{E}(t) \cos(\omega_L t), \tag{3}$$

where $\boldsymbol{\mu}$ is dipole moment operator of the molecule, $\boldsymbol{E}(t)$ is the laser field amplitude including the photon polarization unit vector, and ω_L is the central frequency of the laser.

The temporal behavior of the molecule of interest is obtained by solving the time-dependent Schrödinger equation, which is expressed in the Schrödinger picture as

$$i\hbar \frac{\partial}{\partial t} \Psi(t) = [H_0 + V(t)]\Psi(t). \tag{4}$$

It is convenient to transform the above equation into the time-dependent Schrödinger equation in the interaction picture by introducing a unitary operator

$$U_0(t, t_0) = \exp\left[-\frac{iH_0}{\hbar}(t - t_0)\right]. \tag{5}$$

The wavefunction in the interaction picture $\Psi^I(t)$ is related to that in the Schrödinger picture $\Psi(t)$ through the unitary operator as

$$\Psi(t) = U_0(t, t_0)\Psi^I(t) \quad \text{with} \quad \Psi(t_0) = \Psi^I(t_0). \tag{6}$$

The time-dependent Schrödinger equation in the interaction picture is expressed as

$$i\hbar \frac{\partial}{\partial t} \Psi^I(t) = V^I(t) \Psi^I(t), \tag{7}$$

where $V^I(t)$, the molecule–laser field interaction operator in the interaction picture, is given as

$$V^I(t) = U_0(t, t_0)^\dagger V(t) U_0(t, t_0). \tag{8}$$

The formal solution of Eq. (7) is expressed in the expansion form as

$$\Psi^I(t) = \hat{T} \exp\left[-\frac{i}{\hbar} \int_{t_0}^{t} V^I(t') dt'\right] \Psi^I(t_0). \tag{9}$$

Here, $\hat{T}$, called a time-ordering operator, is defined as

$$\begin{cases} \hat{T}\{V^I(t_1)V^I(t_2)\} = V^I(t_1)V^I(t_2) & \text{for } t_1 > t_2 \\ \hat{T}\{V^I(t_1)V^I(t_2)\} = V^I(t_2)V^I(t_1) & \text{for } t_2 > t_1 \end{cases}. \tag{10}$$

The wavefunction in the Schrödinger picture is given after substituting Eq. (9) to Eq. (6) as

$$\Psi(t) = \exp\left[-\frac{iH_0}{\hbar}(t - t_0)\right] \hat{T} \exp\left[-\frac{i}{\hbar} \int_{t_0}^{t} V^I(t') dt'\right] \Psi(t_0). \tag{11}$$

If $V(t) = 0$, the solution is simply written as

$$\Psi(t) = U_0(t, t_0)\Psi(t_0) = \exp\left[-\frac{iH_0}{\hbar}(t - t_0)\right]\Psi(t_0). \quad (12)$$

The time evolution operator $\exp\left[-\frac{iH_0}{\hbar}(t - t_0)\right]$ represents the free propagator that shifts the state Ψ from time t_0 to t.

Let us now look through how a nuclear wavepacket is generated by an optical excitation. For this purpose, it is instructive to adopt a two-electronic state model in a weak laser field case. In this case, a treatment of the first-order time-dependent perturbation is sufficient. The initial wavefunction is expressed within the Born–Oppenheimer approximation as a product of electronic wavefunction $\Phi_g(r, R)$ and nuclear wavefunction $X_{gv}(R)$. We omit temperature effects for simplicity and consider a low temperature limit case.

Within the Born–Oppenheimer approximation, the nuclear wavefunction generated on e excited-state potential at time t, $|X_e(t)\rangle$, is obtained by projecting Eq. (11) to the electronic state $\Phi_e(r, R)$ as

$$|X_e(t)\rangle = \langle\Phi_e|\Psi(t)\rangle_r$$

$$\simeq \left\langle\Phi_e\left|\exp\left[-\frac{iH_0}{\hbar}(t - t_0)\right]\left[1 - \frac{i}{\hbar}\int_{t_0}^t V^I(t')dt'\right]\Phi_g X_{gv}\right\rangle_r, \quad (13)$$

where $\langle\rangle_r$ denotes the integration over the electronic coordinates of the molecular system. Using the orthogonality of the electronic wavefunctions, $\langle\Phi_e|\Phi_g\rangle_r = 0$, and noting that $H_0|\Phi_e(r, R)\rangle = h_e|\Phi_e(r, R)\rangle$, with $h_e(=T_R + U_e(R))$, the nuclear Hamiltonians of the electronic excited state e, $U_e(R)$ the excited-state potential, and $H_0|\Phi_g(r, R)X_{g0}\rangle = E_{g0}|\Phi_g(r, R)X_{g0}\rangle$ with the lowest vibrational eigenstate in the electronic ground state $|X_{g0}\rangle$, Eq. (13) can be expressed as,

$$|X_e(t)\rangle \simeq \frac{i}{\hbar}\int_{t_0}^t \exp\left[-\frac{ih_e}{\hbar}(t - t')\right]\mu_{eg}(R) \cdot E(t')\exp\left[-\frac{iE_{g0}}{\hbar}t'\right]dt'$$

$$\times |X_{g0}\rangle. \quad (14)$$

In Eq. (14), $\mu_{eg}(R)(= \langle\Phi_e|\mu|\Phi_g\rangle_r)$ is the electronic transition moment vector at the nuclear configuration R.

For simplicity, we adopt a delta function excitation $E(t') = E_0\delta(t')$ in which E_0 the amplitude of the electric field of laser with photon polarization and take $t_0 \to -\infty$. Equation (14) is expressed as

$$|X_e(t)\rangle \simeq \frac{i}{\hbar}\exp\left[-\frac{ih_e}{\hbar}t\right]\mu_{eg}(R) \cdot E_0|X_{g0}\rangle$$

$$= \exp\left[-\frac{ih_e}{\hbar}t\right]|X_e(0)\rangle. \quad (15)$$

Here, $|X_e(0)\rangle$ is the wavepacket generated on the excited-state potential by the delta function excitation, and has the form

$$|X_e(0)\rangle = \frac{i}{\hbar}\boldsymbol{\mu}_{eg}(R)\cdot\boldsymbol{E}_0|X_{g0}\rangle. \tag{16}$$

Here, $|X_{g0}\rangle$ is the eigenstate of the lowest vibrational state in the ground state. Equation (16) shows that the nuclear wavefunction generated by the delta function excitation has a similar shape to that in the electronic ground state. That is, the nuclear wavefunction is localized on the excited-state potential as shown in Fig. 1 of Chap. 1. The nuclear wavefunction $|X_e(t)\rangle$, which is the nuclear wavepacket at time t, freely propagates on the e excited-state potential after optical excitation as shown in Fig. 1 of Chap. 1.

To see the characteristic features of the nuclear wavepacket, we expand the wavepacket in terms of the vibrational eigenfunctions in the excited state. The time-independent Schrödinger equation for the nuclear motion in the electronic state e is given as

$$[T_R + U_e(R)]|X_{en}\rangle = E_{en}|X_{en}\rangle. \tag{17}$$

Here, $|X_{en}\rangle$, the nth vibrational eigenfunction with eigenvalue E_{en}, satisfies the completeness relation, i.e.,

$$\sum_n |X_{en}\rangle\langle X_{en}| = \mathbf{1}. \tag{18}$$

Here, $\mathbf{1}$ denotes the unit operator.

Applying the completeness relation Eq. (18) to (16), the wavepacket is expressed as a linear combination (coherent superposition) of the nuclear eigenfunctions as

$$|X_e(0)\rangle = \frac{i}{\hbar}\boldsymbol{\mu}_{eg}\cdot\boldsymbol{E}_0\sum_n S_{en,g0}|X_{en}\rangle, \tag{19}$$

with weighting factor $S_{en,g0}(=\langle X_{en}|X_{g0}\rangle)$, which is called the optical Franck–Condon overlap integral between the lowest ground state $|g0\rangle$ and vibronic state $|en\rangle$. In Eq. (19), the R-dependence in the transition moment has been omitted for simplicity.

The nuclear wavepacket at time t is expanded as

$$|X_e(t)\rangle \simeq \frac{i}{\hbar}\boldsymbol{\mu}_{eg}\cdot\boldsymbol{E}_0\sum_n S_{en,g0}\exp\left[-\frac{iE_{en}}{\hbar}t\right]|X_{en}\rangle. \tag{20}$$

So far a delta function optical excitation was assumed in generating nuclear wavepackets. For an optical excitation whose interaction is given in Eq. (3), the nuclear wavepacket has the form

$$|X_e(t)\rangle \simeq \frac{i}{2\hbar} \int_{t_0}^{t} \exp\left[-\frac{ih_e}{\hbar}(t-t')\right] \boldsymbol{\mu}_{eg}(R) \cdot \boldsymbol{E}(t')\{\exp(-i\omega_L t')$$
$$+ \exp(i\omega_L t')\} \exp\left[-\frac{iE_{g0}}{\hbar}t'\right] dt' |X_{g0}\rangle. \qquad (21)$$

In the rotating wave approximation, Eq. (21) can be expressed omitting the nonresonant term as

$$|X_e(t)\rangle \simeq \frac{i}{2\hbar} \exp\left[-\frac{ih_e}{\hbar}t\right] \int_{t_0}^{t} \boldsymbol{\mu}_{eg}(R) \cdot \boldsymbol{E}(t')$$
$$\times \exp\left[\frac{i(h_e - E_{g0} - \hbar\omega_L)}{\hbar}t'\right] dt' |X_{g0}\rangle. \qquad (22)$$

For a pulse excitation with its duration $0 \sim t_1$, the wavepacket is expressed as:

$$|X_e(t)\rangle \simeq \exp\left[-\frac{ih_e}{\hbar}(t-t_1)\right] |X_e(t_1)\rangle, \qquad (23)$$

with the nuclear wavepacket at time t_1, which is written as

$$|X_e(t_1)\rangle = \frac{i}{2\hbar} \exp\left[-\frac{ih_e}{\hbar}t_1\right] \int_0^{t_1} \boldsymbol{\mu}_{eg}(R) \cdot \boldsymbol{E}(t')$$
$$\times \exp\left[\frac{i(h_e - E_{g0} - \hbar\omega_L)}{\hbar}t'\right] dt' |X_{g0}\rangle. \qquad (24)$$

The generation of wavepackets can be proved directly by observing beating signals, called quantum beats, in time-resolved spectra. The creation of quantum beats is the direct consequence of the coherent superposition of vibronic states, which interfere with each other.

2. Numerical Methods for Wavepacket Propagation

In the preceding section, the behaviors of nuclear wavepackets in a weak laser field were introduced within the first-order perturbation with respect to molecule–laser interactions. In order to describe the time evolution of electronic and/or nuclear wavepackets generated in an experiment, it is necessary to use a quantitative method beyond the finite perturbation method that is applicable not only to weak laser field but also to intense laser field cases. The time-dependent Schrödinger equation of a molecular system in intense laser fields must be numerically solved. In this section, several numerical methods for time evolution of wavepackets are introduced.

2.1 Symmetrized split operator method

The formal solution of the time-dependent Schrödinger equation (4) is given in terms of unitary operator $U(t, t_0)$ as

$$\psi(t) = U(t, t_0)\psi(t_0). \quad (25)$$

Here,

$$U(t, t_0) = \hat{T} \exp\left[-\frac{i}{\hbar} \int_{t_0}^{t} H(t')dt'\right], \quad (26)$$

with $H(t')$ given by Eq. (1).

The unitary operator $U(t, t_0)$ has two types of noncommuting operators, the momentum and position operators of nucleus and electrons in the exponent. To solve $U(t, t_0)$, we adopt a practical numerical method that is based on the Trotter production formula[3] and is often called the split operator method.[4] The method is effective in a short time range.

When two Hermitian operators A and B are noncommuting, $AB - BA \neq 0$, unitary operator $\exp[(A + B)\Delta t]$ can be generally expanded as

$$\exp[(A + B)\Delta t] = \lim_{n \to \infty} \left(\exp\left[\frac{A\Delta t}{n}\right]\exp\left[\frac{B\Delta t}{n}\right]\right)^n. \quad (27)$$

The symmetrized Trotter formula is given as an approximate expression for the unitary operator by restricting $n = 2$ in Eq. (27) as

$$\exp[(A + B)\Delta t] \simeq \exp\left[\frac{B\Delta t}{2}\right]\exp[A\Delta t]\exp\left[\frac{B\Delta t}{2}\right] + O(\Delta t^3). \quad (28)$$

The error of the symmetrized split operator method is in the order of $(\Delta t)^3$. The norm of the wavefunction is conserved.

As an example, we apply the symmetrized split operator to a time evolution of a 1D system of a molecule with mass m, which is characterized by a time-independent Hamiltonian H with the initial wavepacket $\psi(x, t_0)$ at time $t = t_0$ as

$$H = -\frac{\hbar^2}{2m}\frac{d^2}{dx^2} + V(x). \quad (29)$$

Here, $V(x)$ is the potential energy. The wavefunction at time $t_0 + \Delta t$ can be obtained by operating the time evolution operator to the wavefunction at t_0 as

$$\psi(x, t_0 + \Delta t) = \exp\left[-i\frac{\Delta t}{\hbar}\left(-\frac{\hbar^2}{2m}\frac{d^2}{dx^2} + V(x)\right)\right]\psi(x, t_0). \quad (30)$$

Here, $\psi(x, t_0)$ is the initial wavepacket at $t = t_0$. If Δt is sufficiently small, the wavefunction can be expanded in terms of the symmetrized split operator as

$$\psi(x, t_0 + \Delta t) \simeq \exp\left[-i\frac{V(x)\Delta t}{2\hbar}\right] \exp\left[i\frac{\hbar \Delta t}{2m}\frac{d^2}{dx^2}\right] \exp\left[-i\frac{V(x)\Delta t}{2\hbar}\right] \times \psi(x, t_0). \tag{31}$$

This expansion is valid to the second order with respect to Δt. It can be seen that the norm is conserved since

$$\int |\psi(x, t_0 + \Delta t)|^2 dx$$

$$= \int \psi^*(x, t_0) \exp\left[i\frac{V(x)\Delta t}{2\hbar}\right] \exp\left[-i\frac{\hbar \Delta t}{2m}\frac{d^2}{dx^2}\right] \exp\left[i\frac{V(x)\Delta t}{2\hbar}\right]$$

$$\times \exp\left[-i\frac{V(x)\Delta t}{2\hbar}\right] \exp\left[i\frac{\hbar \Delta t}{2m}\frac{d^2}{dx^2}\right] \exp\left[-i\frac{V(x)\Delta t}{2\hbar}\right] \psi(x, t_0) dx$$

$$= \int |\psi(x, t_0)|^2 dx. \tag{32}$$

Computation of the wavepacket propagation consists of multiplying $\psi(r, t_0)$ by three exponential operators. In the first multiplication, the wavepacket $\psi(r, t_0)$ expressed in the coordinate space is simply multiplied by $\exp\left[-i\frac{V(r)\Delta t}{2\hbar}\right]$ since this operator is also expressed in the coordinate representation. In the next multiplication, the resultant wavepacket $\exp\left[-i\frac{V(r)\Delta t}{2\hbar}\right]\psi(r, t_0)$ is first transformed into the momentum space by using an inverse Fourier transform. The resultant expression is then multiplied by $\exp\left[i\frac{\hbar \Delta t}{2m}\frac{d^2}{dr^2}\right]$. In the third multiplication, the wavepacket is transformed back into the coordinate space and multiplied by the remaining exponential operator, $\exp\left[-i\frac{V(r)\Delta t}{2\hbar}\right]$. Sequential multiplication of the three exponential operators in Eq. (31) can be efficiently carried out by using the fast Fourier transform (FFT) algorithm.

2.2 Finite difference methods

The essence of the finite difference methods is replacement of the derivatives in (time-dependent) Schrödinger equations by appropriate finite differences. Time and space of the system are represented in terms of grids. Electrons in atoms and molecules move in Coulomb potential, which is characterized by its long-range force and its singularity at the origin, and a special care must be taken compared with nuclear motions. This means that the boundary has to be chosen to be long enough and the grid spacing has to be small to adjust to the steepness of the Coulomb potential near the origin. The split operator method, which is an efficient method for nuclear systems, has been found to be inefficient for systems characterized by a Coulomb potential with singularity.[5]

Let us consider a 1D molecular system with mass m interacting with a laser field. The time-dependent Schrödinger equation is given in the semiclassical treatment as

$$i\hbar \frac{\partial \psi}{\partial t} = H\psi, \tag{33}$$

where $H = K + V$. Kinetic energy operator K is given as $K = -\frac{\hbar^2}{2m}\frac{d^2}{dx^2}$, and $V(= V(x, t))$ involves both the potential energy and time-dependent molecule–laser field interaction. The wavefunction, $\psi(= \psi(x, t))$, satisfies the two boundary conditions: $\psi(x = x_0, t) = 0$ and $\psi(x = x_f, t) = 0$, where the grid end x_f is chosen so that the amplitude of the wavefunction is negligibly small for $x > x_{\max}$. The range $[0, x_f]$ is discretized by N grid points. In the finite difference method, ψ_j^n is expressed by its values at a discrete set of points of both 1D space x and time t as $x_j = x_0 + j\Delta x$, with $j = 0, 1, \ldots, j_{\max}$ and $t_n = t_0 + n\Delta t$, with $n = 0, 1, \ldots, n_{\max}$, respectively.

Let the time derivative be expressed in terms of a two-point formula and let the second derivative in the kinetic operator K be approximated in terms of the ordinary three-point finite difference. The time-dependent Schrödinger equation is expressed by using the finite difference method[6] as

$$i\hbar \left(\frac{\psi^{n+1} - \psi^n}{\Delta t}\right) = H^{n+\frac{1}{2}} \psi^{n+\frac{1}{2}}, \tag{34a}$$

or

$$i\hbar \left(\frac{\psi_j^{n+1} - \psi_j^n}{\Delta t}\right) = \sum_k H_{jk}^{n+\frac{1}{2}} \psi_k^{n+\frac{1}{2}}. \tag{34b}$$

Here,

$$H_{jk}^{n+\frac{1}{2}} = [K + V(r, t)]_{jk}^{n+\frac{1}{2}}$$

$$= -\frac{\hbar^2}{2m(\Delta x)^2}(\delta_{jk+1} - 2\delta_{jk} + \delta_{jk-1}) + V_j^{n+\frac{1}{2}} \delta_{jk}, \tag{35}$$

where the Hamiltonian matrix of the system is tridiagonal, i.e., the nonzero matrix elements are on or adjacent to the diagonal one.

In Eq. (35), $H_{jk}^{n+\frac{1}{2}}$ is the Hamiltonian matrix element at the half time-step

$$t_{n+\frac{1}{2}} = t_0 + \left(n + \frac{1}{2}\right)\Delta t.$$

Note that the right-hand side of Eq. (34a) requires knowledge of $(H\psi)^{n+\frac{1}{2}}$ at the half time-point. Since the wavefunction at the half time-point is unknown, $\psi^{n+\frac{1}{2}}$ is approximated by averaging between ψ^n and ψ^{n+1}. Thus, Eq. (34a) is given as

$$i\hbar \left(\frac{\psi^{n+1} - \psi^n}{\Delta t} \right) = \frac{1}{2} H^{n+\frac{1}{2}} (\psi^n + \psi^{n+1}). \tag{36}$$

This equation can be rewritten as

$$\left(1 + i \frac{H^{n+\frac{1}{2}} \Delta t}{2\hbar} \right) \psi^{n+1} = \left(1 - i \frac{H^{n+\frac{1}{2}} \Delta t}{2\hbar} \right) \psi^n. \tag{37}$$

The above equation has a solution called the Cayley form as

$$\psi^{n+1} = L^{n+\frac{1}{2}} \psi^n, \tag{38}$$

where

$$L^{n+\frac{1}{2}} = \left(1 + i \frac{H^{n+\frac{1}{2}} \Delta t}{2\hbar} \right)^{-1} \left(1 - i \frac{H^{n+\frac{1}{2}} \Delta t}{2\hbar} \right). \tag{39}$$

This differencing scheme, which is second-order accurate in time, is called the Cayley–Crank–Nicholson (CCN) scheme.[7]

The operator $L^{n+\frac{1}{2}}$ is unitary and the norm of the wavefunction is conserved as shown below.

Let $L^{n+\frac{1}{2}\dagger}$ be the adjoint of $L^{n+\frac{1}{2}}$:

$$L^{n+\frac{1}{2}\dagger} = \left(1 + i \frac{H^{n+\frac{1}{2}} \Delta t}{2\hbar} \right) \left(1 - i \frac{H^{n+\frac{1}{2}} \Delta t}{2\hbar} \right)^{-1}. \tag{40}$$

It can easily be shown that $L^{n+\frac{1}{2}} L^{n+\frac{1}{2}\dagger} = 1$ since $\left(1 + i \frac{H^{n+\frac{1}{2}} \Delta t}{2\hbar} \right)$ and $\left(1 - i \frac{H^{n+\frac{1}{2}} \Delta t}{2\hbar} \right)$ are commuting and $H = H^{\dagger}$, and $L^{n+\frac{1}{2}}$ is unitary.

The norm can be expressed as

$$\langle \psi^{n+1} | \psi^{n+1} \rangle = \langle L^{n+\frac{1}{2}} \psi^n | L^{n+\frac{1}{2}} \psi^n \rangle = \langle \psi^n | \psi^n \rangle. \tag{41}$$

This means that the norm is time-independent and conserved.

Although the CCN scheme described above is second-order accurate in time and the norm is conserved, the efficiency in computation for a multi-dimensional system is rather poor. To circumvent the inefficiency in computation, there is another method called the alternating-direction implicit (ADI) method,[8] which is a generalized version of the Cayley–Crank–Nicholson scheme. Time splitting or the method of fractional steps is adopted in the ADI method. The simplest step is the intermediate step $n + 1/2$ to get from n to $n + 1$.

As an example, the time-evolution operator $\exp[-i(A+B)\Delta t]$ in which A and B are arbitrary operators can be approximated as

$$\exp[-i(A+B)\Delta t] \simeq \left(1 + iA\frac{\Delta t}{2}\right)^{-1} \frac{\left(1 - iB\frac{\Delta t}{2}\right)}{\left(1 + iB\frac{\Delta t}{2}\right)} \left(1 - iA\frac{\Delta t}{2}\right). \quad (42)$$

The above approximated operator consists of two sequential Cayley–Crank–Nicholson (CCN) schemes. The operation can conveniently be carried out by introducing an intermediate state $\psi^{n+\frac{1}{2}}$ and solving two implicit equations:

$$\left(1 + iB\frac{\Delta t}{2}\right)\psi^{n+\frac{1}{2}} = \left(1 - iA\frac{\Delta t}{2}\right)\psi^n \quad (43a)$$

and

$$\left(1 + iA\frac{\Delta t}{2}\right)\psi^{n+1} = \left(1 - iB\frac{\Delta t}{2}\right)\psi^{n+\frac{1}{2}}, \quad (43b)$$

which is known as the Peaceman–Rachford (PR) method.[9]

The result is given as

$$\begin{aligned}\exp[-i(A+B)\Delta t]\psi^n &\simeq \left(1 + iA\frac{\Delta t}{2}\right)^{-1} \frac{\left(1 - iB\frac{\Delta t}{2}\right)}{\left(1 + iB\frac{\Delta t}{2}\right)} \left(1 - iA\frac{\Delta t}{2}\right)\psi^n \\ &\simeq \left(1 + iA\frac{\Delta t}{2}\right)^{-1} \left(1 - iB\frac{\Delta t}{2}\right)\psi^{n+\frac{1}{2}} \\ &\simeq \psi^{n+1}. \end{aligned} \quad (44)$$

The CCN scheme or the scheme in the PR method is an "implicit" evaluation scheme since, for example, in order to calculate $\left(1 + iB\frac{\Delta t}{2}\right)\psi^{n+\frac{1}{2}} = \left(1 - iA\frac{\Delta t}{2}\right)\psi^n$, operator $\left(1 + iB\frac{\Delta t}{2}\right)$ needs to be inverted. On the other hand, the equation of the type $\psi^{n+1} = (1 + iA\Delta t)\psi^n$, for example, is solved in an "explicit" scheme since the matrix involving the operator at any time step is calculated in terms of the known values of the matrix at past times without the need to evaluate a system of equations.

The two-step scheme in the PR method given by Eq. (43) is useful for a 2D system. As an example, consider a molecular system in which operator A contains differential operators with respect to x and B contains those of y. The three-point finite differential formula in the CCN scheme can be applied to the operation of differential operators on the wavefunctions. Equation (42) is reduced to two sets of trigonal equations by using the PR method: using Eq. (43a), a tridiagonal system of equations connecting the probability amplitudes at grid points $\{y_i\}$ is obtained at each x grid point, and using Eq. (43b), a tridiagonal system of equations connecting the probability amplitudes at $\{x_i\}$ is obtained at each y. The whole system is reduced to two sets of subsystems in one space dimension. The tridiagonal systems of equations can be solved by setting the wavefunction to zero at appropriate boundaries.

The alternating-direction implicit (ADI) method can be extended to 3D systems in a similar way. Let us consider a 3D system in which differential operators of three degrees of freedom are involved in A, B, and C, respectively. The propagator can be expressed as

$$\exp[-i(A + B + C)\Delta t]$$
$$\simeq \left(1 + iA\frac{\Delta t}{2}\right)^{-1}\left(1 + iB\frac{\Delta t}{2}\right)^{-1}\frac{\left(1 - iC\frac{\Delta t}{2}\right)}{\left(1 + iC\frac{\Delta t}{2}\right)}\left(1 - iB\frac{\Delta t}{2}\right)\left(1 - iA\frac{\Delta t}{2}\right). \tag{45}$$

The propagation from ψ^n to ψ^{n+1} is carried out by three steps *via* two intermediate states $\psi^{n+\frac{1}{3}}$ and $\psi^{n+\frac{2}{3}}$ as

$$\left(1 + iC\frac{\Delta t}{2}\right)\psi^{n+\frac{1}{3}} = \left(1 - iC\frac{\Delta t}{2}\right)\left(1 - iB\frac{\Delta t}{2}\right)\left(1 - iA\frac{\Delta t}{2}\right)\psi^n, \tag{46a}$$

$$\left(1 + iB\frac{\Delta t}{2}\right)\psi^{n+\frac{2}{3}} = \psi^{n+\frac{1}{3}}, \tag{46b}$$

and

$$\left(1 + iA\frac{\Delta t}{2}\right)\psi^{n+1} = \psi^{n+\frac{2}{3}}. \tag{46c}$$

Thus, the wavefunction ψ^{n+1} is obtained by solving Eq. (46a) to Eq. (46c) in order.

2.3 Dual transformation technique

The dual transformation technique is an efficient grid method for describing the electronic dynamics in atoms and molecules.[10] The electronic dynamics is characterized by long-range Coulomb potential and its singularity at the origin. In this technique, both the wavefunctions and the Hamiltonian of the molecule of interest are consistently transformed for the electronic system to satisfy the time-dependent Schrödinger equation in the grid scheme. The coordinate systems should be chosen to describe electronic dynamics. Three conditions should be satisfied to describe correctly the electronic dynamics: (i) the wavefunction is transformed to be zero at the Coulomb singular points in order to avoid the numerical difficulty in treating the singularity, (ii) the differential operators can be evaluated by the finite difference method even near the Coulomb singular points, and (iii) equal spacings in the newly scaled coordinates correspond to grid spacing in the original coordinates that are narrow near the nuclei in order to deal with relatively high momentum components near the nuclei and are wide at larger distances.

As an example of application of dual transformation, let us consider a 3D model of a hydrogen molecular ion in a linearly polarized laser field as shown in Fig. 1 of Chap. 5. The nuclear motion is restricted to the polarization direction z of the electric field of the laser. The cylindrical coordinates are taken to be the original ones. The z-component of the electronic angular momentum is conserved. Therefore, the electronic motions are described by the two cylindrical coordinates z and ρ. The other coordinate is R, which describes the nuclear motion. See Chap. 5 for detailed description of the coordinates.

The time-dependent Schrödinger equation of the relative motions of H_2^+ is given in atomic units as[10]

$$i\frac{\partial}{\partial t}\psi(\rho, z, R) = [T_n(R) + T_e(\rho, z) + U(\rho, z, R) + V_E(z, t)]\psi(\rho, z, R), \quad (47)$$

where $T_n(R)$ is the nuclear kinetic operator, $T_e(\rho, z)$ is the kinetic energy of the electron, $U(\rho, z, R)$ is the Coulomb potential energy, and $V_E(z, t)$ is the interaction between the electron and the electric field of an applied laser.

The original wavefunction that is finite at the nuclei satisfies the normalization condition

$$\int_0^\infty dR \int_0^\infty d\rho \int_{-\infty}^\infty dz \rho |\psi(\rho, z, R)|^2 = 1. \quad (48)$$

It is well known that the norm of the wavefunction is not conserved when the finite differential method is adopted.

Let us now define the generalized cylindrical coordinates $\rho = f(\xi)$ and $z = g(\varsigma)$ in terms of the scaled coordinates ξ and ς as

$$f(\xi) = \xi \left(\frac{\xi^n}{\xi^n + \alpha^n}\right)^\nu \quad \text{and} \quad g(\varsigma) = \left[1 - (1-\beta)\exp\left(-\frac{\varsigma^2}{\gamma^2}\right)\right]\varsigma, \quad (49)$$

where α and γ are widths of ρ- and z-ranges, respectively, and ν and β are parameters.

The generalized cylindrical coordinates lead to the original cylindrical ones when $\nu = 0$ and $\beta = 1$. Let the transformed wavefunction $\psi(\xi, \varsigma, R)$ be satisfied with the normalization condition as

$$\int_0^\infty dR \int_0^\infty d\xi \int_{-\infty}^\infty d\varsigma |\psi(\xi, \varsigma, R)|^2 = 1. \quad (50)$$

The transformed wavefunction has the form[10]

$$\psi(\xi, \varsigma, R) = \sqrt{f(\xi)\frac{df(\xi)}{d\xi}\frac{dg(\varsigma)}{d\varsigma}}\psi(f(\xi), g(\varsigma), R). \quad (51)$$

The wavefunction is zero and analytic at the nuclei, and the equal spacings in the scaled coordinates correspond to grid spacings in the original coordinates, which are narrow near the nuclei and are wide at larger distances. The transformed wavefunction satisfies conditions (i) and (iii), which are required for the wavefunction.

The time-dependent Schrödinger equation is given as

$$i\frac{\partial}{\partial t}\psi(\xi, \varsigma, R) = H(\xi, \varsigma, R)\psi(\xi, \varsigma, R). \tag{52}$$

The transformed Hamiltonian in Eq. (52), $H(\xi, \varsigma, R)$, is expressed as[10]

$$H(\xi, \varsigma, R) = T_R + T_\xi + T_\varsigma + \frac{m^2}{2f^2(\xi)} + U(\xi, \varsigma, R) + V_E(\varsigma, t), \tag{53}$$

where m is the z component of the electronic angular momentum, and the kinetic energy operators T_ξ and T_ς are respectively written as[10]

$$T_\xi = -\frac{1}{4\mu}\left(\frac{1}{\left(\frac{\partial f(\xi)}{\partial \xi}\right)^2}\frac{\partial^2}{\partial \xi^2} + \frac{\partial^2}{\partial \xi^2}\frac{1}{\left(\frac{\partial f(\xi)}{\partial \xi}\right)^2}\right) - \frac{1}{8\mu f^2(\xi)}$$

$$+ \frac{1}{4\mu\left(\frac{\partial f(\xi)}{\partial \xi}\right)^4}\left(\frac{7}{2}\left(\frac{\partial f^2(\xi)}{\partial \xi^2}\right)^2 - \frac{\partial f(\xi)}{\partial \xi}\frac{\partial^3 f(\xi)}{\partial \xi^3}\right), \tag{54a}$$

$$T_\varsigma = -\frac{1}{4\mu}\left(\frac{1}{\left(\frac{\partial g(\varsigma)}{\partial \varsigma}\right)^2}\frac{\partial^2}{\partial \varsigma^2} + \frac{\partial^2}{\partial \varsigma^2}\frac{1}{\left(\frac{\partial g(\varsigma)}{\partial \varsigma}\right)^2}\right)$$

$$+ \frac{1}{4\mu\left(\frac{\partial g(\varsigma)}{\partial \varsigma}\right)^4}\left(\frac{7}{2}\left(\frac{\partial^2 g(\varsigma)}{\partial \varsigma^2}\right)^2 - \frac{\partial g(\varsigma)}{\partial \varsigma}\frac{\partial^3 g(\varsigma)}{\partial \varsigma^3}\right), \tag{54b}$$

and

$$T_R = -\frac{1}{m_p}\frac{\partial^2}{\partial R^2}, \tag{54c}$$

where $\mu = (2m_p m_e)/(2m_p + m_e)$ with proton mass m_p and electron mass m_e.

The prefactor $\sqrt{f(\partial f/\partial \xi)(\partial g/\partial \varsigma)}$ in Eq. (51) changes as an order of $\sqrt{(1+n\nu)\beta}\xi^{\frac{2n\nu+1}{2}}/\alpha^{n\nu}$. From the viewpoint of condition (i) zero value for the wavefunction at the singular point, the exponent of ξ, $\frac{2n\nu+1}{2}$, should be a positive integer.

For the time evolution of the 3D system of H_2^+ in Eq. (45), which is expressed by using the ADI method, the operators A, B, and C are given as

$$A = T_\xi + \frac{1}{2}U(f(\xi), g(\varsigma), R) + V_E\left(g(\varsigma), t_{n+\frac{1}{2}}\right), \tag{55a}$$

$$B = T_\varsigma + \frac{1}{2}U(f(\xi), g(\varsigma), R) + \frac{m^2}{2f^2(\xi)}, \tag{55b}$$

and

$$C = T_R + \frac{1}{R}. \tag{55c}$$

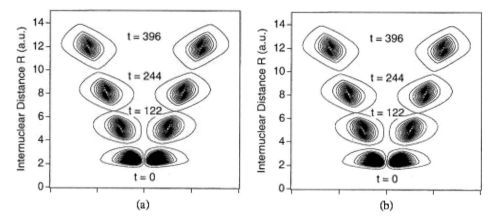

Fig. 1. Contour maps of time-dependent probability on the first electronic excited state of H_2^+. The end of the linearly polarized laser pulse used is set to be the origin of time, $t = 0$. The horizontal axis denotes z. In (a), $n = 1$ and $\nu = 1/2$ are adopted. In (b), $n = 1/2$ and $\nu = 1$ are adopted. The wavepacket form at $t = 0$ has a node, which indicates that the initial wavepacket is created on the first excited electronic state of the $1\sigma_u$ antibonding MO character. (Reproduction with permission from Kawata and Kono.[10] Copyright (1999) by American Institute of Physics.)

In Eqs. (55a) and (55b), the Coulomb potential term, U, is divided into halves, so that $A\psi$ and $B\psi$ vanish at the nuclei. This guarantees that intermediates $\psi^{n+\frac{1}{3}}$ and $\psi^{n+\frac{2}{3}}$ as well as $\psi(t_n)$ give zero at the nuclei.[10] That is, the propagator with the form described above satisfies condition (ii) for the proper wavefunction.

Figure 1 shows contour maps of the time-dependent probability of H_2^+ as a function of z and R, which has the form $\int |\psi(\rho, z, R)|^2 \rho d\rho$.

The electric field of the laser used in the calculation is given as

$$E(t) = \sin\frac{\pi(t+T)}{T} \sin\omega(t+T)$$

$$\text{for } -T \leq t \leq 0, \text{ otherwise } E(t) = 0,$$

where $T = 100$ (2.5 fs) as the pulse duration, and ω, the central frequency of the laser, is set to be equal to the energy difference between $1\sigma_g$ and $1\sigma_u$ at the equilibrium internuclear distance $R = 2.0$. The grid ends are chosen as $\rho_{max} = 8.83$ and $z_{max} = 10$. $\alpha = 28.3$ and $\Delta R = 0.05$ are set as the parameters in the scaled coordinates. Figure 1 shows the ultrafast excitation and dissociation dynamics of H_2^+. The wavepacket propagation shown in Fig. 1(a) can be regarded as the exact one since the numbers of grid points are large enough to describe the time evolution of the wavepacket: $N_\xi = 151$ and $N_\zeta = 207$ ($\Delta\xi = 0.1$ and $\Delta\zeta = 0.1$). The numbers of grid points are reduced to $N_\xi = 19$ and $N_\zeta = 57$ in Fig. 1(b). It can be seen that the global features of the wavepacket propagations in Fig. 1(b) are reproduced well even with the small number of grid points used.

3. Wavepacket Propagation Method in the Scattering Matrix Framework

The scattering matrix (S-matrix) method is a general method for evaluation of the transition probability from an initial to a continuum state.[11] Theory based on the S-matrix method for calculation of ionization rates of atoms in intense laser fields is known as the Keldysh–Faisal–Reiss (KFR) theory.[12] It was extended to many-body problems of atoms and molecules in intense laser fields,[13] which is sometimes called the intense-field many-body S-matrix theory. For molecules, intramolecular electronic and nuclear dynamics as well as ionization occur in intense laser fields as can be seen in Chap. 5.

In this section, a practical method for taking into account the dynamics competing with ionization is briefly described.[14] The method is a combination of the S-matrix and the wavepacket method by taking advantage of their merits. Series expansion of the transition amplitude in the ordinary S-matrix method can be rearranged systematically by choosing the essential processes, while the wavepacket propagation approach is nonperturbative by directly solving the equations of motion for intramolecular dynamics. The wavepacket composed of only bound electronic states is introduced into a framework of the intense field S-matrix.

The S-matrix element for the transition from an initial state, $|\psi_i(t_0)\rangle$, to a final continuum state, $|\psi_f(t = \infty)\rangle$, is given as

$$S_{fi} = \lim_{t \to \infty} \langle \psi_f(t) | \Psi_i(t) \rangle, \qquad (56)$$

where $|\Psi_i(t)\rangle$ is the solution of the time-dependent Schrödinger equation

$$i\hbar \frac{\partial}{\partial t} |\Psi_i(t)\rangle = H(t) |\Psi_i(t)\rangle \qquad (57)$$

with the initial condition $|\Psi(t_0)\rangle = |\psi_i(t_0)\rangle$. In Eq. (57), $H(t)$ is the time-dependent Hamiltonian in the semiclassical treatment of molecular–laser interactions. The Hamiltonian can be expressed in terms of the reference Hamiltonians for the initial and final states as

$$H(t) = H_i(t) + H'_i(t) = H_f(t) + H'_f(t). \qquad (58)$$

Here, $H_i(t)$ is the reference Hamiltonian for an initial state and $H_f(t)$ is that for a chosen final state. $H'_i(t)$ and $H'_f(t)$ are the interaction Hamiltonians of the relevant states.

The time-dependent Schrödinger equation (57) is formally solved in the "post" form as

$$|\Psi_i(t)\rangle = U_f(t,t_0)|\Psi_i(t_0)\rangle - \frac{i}{\hbar}\int_{t_0}^t d\tau\, U_f(t,\tau) H'_f(t)|\Psi_i(\tau)\rangle, \quad (59)$$

where $U_f(t,t_0)$ is the unitary operator for the $H_f(t)$. The unitary operator satisfies

$$i\hbar\frac{\partial}{\partial t}U_f(t,t_0) = H_f(t)U_f(t,t_0) \quad (60)$$

with the initial condition $U_f(t_0,t_0) = 1$.

Let us now consider an ionization of a molecule in an intense laser field. The reference Hamiltonian, $H_f(t)$, is the sum of the Hamiltonian for a free electron in the laser field and that for the remaining molecular ion in the laser field. The interaction Hamiltonian $H'_f(t)$ is the attractive interaction between the free electron and the remaining molecular ion. The reference Hamiltonian for the initial state, $H_i(t)$, is the neutral–molecular Hamiltonian and $H'_i(t)$ is the molecule–laser field interactions.

We introduce a projection operator, $P_f(t)(\equiv 1 - \sum_b |\psi_b(t)\rangle\langle\psi_b(t)|)$, which extracts a continuum state of interest at time t. Here, $\{|\psi_b(t)\rangle\}$ denote the bound states. The scattering matrix element between an initial state and the continuum state can be expressed as

$$S_{fi} = \lim_{t\to\infty}\langle\psi_f(t)|P_f(t)|\Psi_i(t)\rangle$$

$$= \lim_{t\to\infty}\langle\psi_f(t)|\left[|\Psi_i(t)\rangle - \sum_b |\psi_b(t)\rangle\langle\psi_b(t)|\Psi_i(t)\rangle\right]. \quad (61)$$

Here, $\langle\psi_b(t)|\Psi_i(t)\rangle$ denotes the amplitude of bth bound-state involved in $|\Psi_i(t)\rangle$, $c_{ib}(t)(\equiv \langle\psi_b(t)|\Psi_i(t)\rangle)$.

Let us simply assume that $|\Psi_i(t)\rangle$ involves only the neural molecular components to take into account intramolecular dynamics. Then, the second term of the right-hand side in Eq. (61) can be written as

$$|\Phi_i(t)\rangle = \sum_b c_{ib}(t)|\psi_b(t)\rangle, \quad (62)$$

where $|\Phi_i(t)\rangle$ denotes the wavepacket of the bound states in the laser field, and coefficient $c_{ib}(t)$ satisfies the coupled equation of motion,[14]

$$i\hbar\frac{\partial}{\partial t}c_{ib}(t) = \sum_{b'} c_{ib'}(t)\langle\psi_b(t)|H'_i(t)|\psi_{b'}(t)\rangle. \quad (63)$$

The first term of the right-hand side in Eq. (61) is a solution of Eq. (59) in which $|\Psi_i(\tau)\rangle$ is replaced by $|\Phi_i(\tau)\rangle$. The resultant wavefunction for $|\Psi_i(t)\rangle$ is denoted by

$|\tilde{\Psi}_i(t)\rangle$. The final state, $|\Psi_f(t \to \infty)\rangle$, is approximated by a product of Volkov state and molecular ion core. Finally, the S-matrix element is written as

$$S_{fi} = \lim_{t \to \infty} \langle \phi_f(t)| \left[|\tilde{\Psi}_i(t)\rangle - |\Phi_i(t)\rangle \right]$$

$$= \lim_{t_0 \to -\infty, t \to \infty} \left[\langle \phi_f(t_0)|\psi_i(t_0)\rangle \right.$$

$$\left. - \frac{i}{\hbar} \int_{t_0}^{t} d\tau \langle \phi_f(\tau)|H'_f|\Phi_i(\tau)\rangle - \langle \phi_f(t)|\Phi_i(t)\rangle \right], \qquad (64)$$

where $|\phi_f(t)\rangle$ denotes the Volkov state, which expresses a free electron in a laser field.

The total ionization probability W is given by performing integration of free-electron canonical momenta $\{\boldsymbol{p}_f\}$ over all the ionizing states $\{f\}$ as[14]

$$W = \frac{V}{(2\pi)^3} \int d\boldsymbol{p}_f |S_{fi}|^2. \qquad (65)$$

Here, V is a configuration space volume for normalization of the final states and $(2\pi)^3$ is the unit-volume size in phase space.

Figure 2 shows the results of the S-matrix method applied to ionization of H_2^+ in an intense laser field.[14]

The laser used was assumed to be of a sine-square pulse with a linear polarization along the bond, $I_0 = 8.75 \times 10^{13}$ W/cm^2 for intensity, $\lambda = 760$ nm for central frequency, and $T = 13.3$ fs for duration. The ionization probability is plotted as a function of the internuclear distance R. The magnitudes calculated in the S-matrix method were multiplied by a factor of 0.063, which were irrespective of the values of R. For comparison, results obtained by an accurate grid propagation method, which is described in Sec. 3.3, are indicated by circles. The results (squares) obtained by the S-matrix method with a full expansion in a restricted molecular basis set reproduce the qualitative features of enhanced ionization, *i.e.*, two maxima in ionization probability appear at around $6.0a_0$ and $10a_0$. This reflects enhanced ionization processes from the populated $2p\sigma_u$ state of H_2^+.[15,16] Thus, the present method provides a practical way of inclusion of the effects of intramolecular electronic dynamics. However, quantitatively, a deviation exists in the absolute values of the ionization probability between the accurate grid point method and the S-matrix method. The overestimation of the ionization probability in the S-matrix method originates from neglect of any depletion of the bound state population due to ionization. As the next step, the population depletion can be taken into account by introducing tunneling effects for each bound state to Eq. (63).[14]

In this chapter, numerical methods for calculations of electronic and nuclear wavepackets are presented after a brief description of generation and propagation

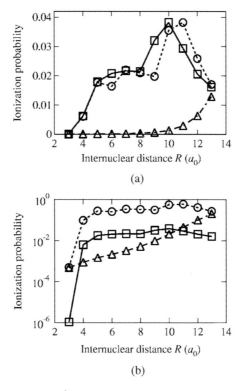

Fig. 2. Ionization probabilities of H_2^+ calculated as a function of the internuclear distance R in atomic units: (a) linear scale plot and (b) logarithmic scale plot. The ionization probabilities calculated by an accurate grid point method are denoted by squares. Two sets for the expansion of $|\Phi_i(t)\rangle$ are shown. The first set is the full expansion in which $|\Phi_i(t)\rangle$ was expanded in terms of all the 48 σ states constructed from the prepared basis set of s and p_z orbitals. In the second set, only the initial $1s\,\sigma_g$ state was used for expansion. The ionization probabilities obtained from the first and second sets are denoted by circles and triangles, respectively. The values obtained from the two sets were multiplied by a factor of 0.063. (Reprinted with permission from Kanno et al.[14] Copyright (2005) by the American Physical Society).

of nuclear wavepackets. Dual transformation technique, which is a grid method, is introduced for describing dynamical system with Coulomb interactions. This technique is used for electronic and nuclear dynamics in atoms and molecules in Chap. 5. Incorporation of a wavepacket propagation method into the S-matrix framework is also presented to take into account the effects of excited state dynamics on intense-field ionization.

References

1. D. J. Tannor, *Introduction to Quantum Mechanics: A Time-Dependent Perspective* (University Science Books, Sausalito, California, 2007).

2. W. H. Louisell, *Quantum Statistical Propertied of Radiation* (John Willey & Sons, New York, 1973).
3. (a) H. F. Trotter, *Proc. Am. Math. Soc.* **10**, 545 (1959); (b) K. E. Schmidt and M. A. Lee, *Phys. Rev. E* **51**, 5495 (1995).
4. (a) J. A. Fleck, Jr., J. R. Morris and M. D. Feit, *Appl. Phys.* **10**, 129, (1976); (b) M. D. Feit and J. A. Fleck, Jr., *J. Chem. Phys.* **80**, 2578 (1984); (c) H. Kono and S. H. Lin, *J. Chem. Phys.* **84**, 1071 (1986); (d) C. Lefoestier, R. H. Bisseling, C. Cerjan, M. D. Feit, R. Friesner, H.-D. Meyer, N. Lipkin, O. Roncero and R. Kosloff, *J. Comp. Phys.* **94**, 59 (1991); (e) T. N. Truong, J. J. Tanner, P. Bala, J. A. McCammon, D. J. Kouri, B. Lesyng and D. K. Hoffman, *J. Chem. Phys.* **96**, 2077 (1992).
5. H. Kono, A. Kita, Y. Ohtsuki and Y. Fujimura, *J. Comp. Phys.* **130**, 148 (1997).
6. P. Bonche, S. Koonin and J. W. Negele, *Phys. Rev. C* **13**, 1226 (1976).
7. (a) W. H. Press, S. A. Teukolsky, W. T. Vetterling and B. P. Flannery, *Numerical Recipes in Fortran*, Second Edition (Cambridge University Press, New York, 1992), Chap. 19; (b) A. Askar and A. S. Cakmak, *J. Chem. Phys.* **68**, 2794 (1978).
8. (a) A. R. Mitchell, *Computational Methods in Partial Differential Equations* (John Wiley & Sons, New York, 1969); (b) N. N. Yanenko, *The Method of Fractional Steps* (Springer-Verlag, New York, 1971).
9. (a) R. S. Varga, *Matrix Iterative Analysis* (Prentice-Hall, Englewood Cliffs, NJ, 1962), p. 273; (b) S. E. Koonin, K. T. R. Davies, V. Maruhn-Rezwani, H. Feldmeier, S. J. Krieger and J. W. Negele, *Phys. Rev. C* **15**, 1359 (1977).
10. I. Kawata and H. Kono, *J. Chem. Phys.* **111**, 9498 (1999).
11. M. L. Goldberger and K. M. Watson, *Collision Theory* (Wiley, New York, 1964).
12. (a) L. V. Keldysh, *Zh. Eksp. Teor. Fiz.* **47**, 1945 (1964) [*Sov. Phys. JETP* **20**, 1307 (1965); (b) F. H. M. Faisal, *J. Phys. B* **6**, L89 (1973); (c) H. R. Reiss, *Phys. Rev. A* **22**, 1786 (1980).
13. (a) F. H. M. Faisal and A. Becker, in *Selected Topics in Electron Physics*, edited by D. M. Campbell and H. Kleinpoppen (Plenum, New York, 1996).
14. M. Kanno, T. Kato, H. Kono, Y. Fujimura and F. H. M. Faisal, *Phys. Rev. A* **72**, 033418 (2005).
15. H. Kono, Y. Sato, Y. Fujimura and I. Kawata, *Laser Phys.* **13**, 883 (2003).
16. H. Kono, Y. Sato, N. Tanaka, T. Kato, K. Nakai, S. Koseki and Y. Fujimura, *Chem. Phys.* **304**, 203 (2004).

Chapter 4

Molecular Manipulation Techniques with Laser Technologies and Their Applications

In this chapter, the molecular alignment and orientation techniques based on laser technologies are outlined first. Then, various applications with a sample of aligned or oriented molecules are reviewed to prove its usefulness. Some basic experimental setups and methods associated with this chapter are described in Chap. 2.

1. Introduction

Since a molecule is generally an anisotropic quantum system, researchers are interested in alignment or orientation dependence in physical and/or chemical processes where molecules are involved.* In addition to the stereodynamics in chemical reactions,[1] alignment and orientation techniques attracted widespread attention in "electronic stereodynamics in molecules."[2] In fact, nonsequential double ionization,[3] high-order harmonic generation,[4–6] and hot above-threshold ionization (ATI) are caused by electron recollision[7] and a sample of aligned or oriented molecules is of crucial importance to investigate these interesting phenomena. It should be noticed that recently remarked molecular orbital tomography is based on high-order harmonic generation from aligned/oriented molecules.[4]

Until the end of 20th century, the molecular orientation techniques are quite limited. Some molecules with exceptionally large permanent dipole moments (a few Debye) can be oriented with a high electrostatic field (a few hundred kV/cm to a few MV/cm), which is called "brute force orientation."[8,9] Symmetric top molecules† in a specific quantum state (J, K, M) can pass through an inhomogeneous electrostatic

---

*"Alignment" means a state where the molecular axes are parallel to each other without any attention paid to the head-versus-tail directions. "Orientation" means a state where the molecular axes are parallel to each other with their heads directed the same way.

†When two of the three principal moments of inertia are equal, a molecule is called a symmetric top.

field created by hexapole electrodes and be focused.[10] They can be further oriented with an electrostatic field. However, there had been no technique to align molecules with no permanent dipole moments.

2. Techniques for Molecular Alignment

2.1 *Adiabatic and nonadiabatic alignment*

The techniques such as brute force orientation and hexapole focusing cannot be applied to nonpolar molecules without permanent dipole moments. Friedrich and Herschbach proposed utilizing the anisotropic polarizability interaction of an intense nonresonant laser field with the induced dipole moment of molecules.[11,12] The interaction creates a double-well potential along the polarization direction of the laser field, which forces the molecules liberate over a limited angular range. The eigenstates of the molecules thus created in the strong laser field are called the pendular states consisting of superpositions of field-free rotational states. Each rotational state with quantum numbers $J|M|$ adiabatically evolves into a particular pendular state by turning on the strong laser field slowly compared to the rotational period of the molecules. This underlying physics tells us that the degree of alignment can be enhanced by using a sample of molecules initially residing in the lowest rotational states because they are transferred into the lowest lying and most aligned pendular states. Alternatively, the degree of alignment can be enhanced by increasing the intensity of the laser field.

Based on the technique described above, Sakai and his coworkers have demonstrated controlling the alignment of neutral I_2 molecules by an intense, nonresonant, and nanosecond laser field.[13] They determine the degree of alignment by photodissociating the molecules during the nanosecond alignment pulse with a femtosecond pump pulse and measuring the directions of the photofragments with a two-dimensional (2D) ion imaging technique.[14] Since the detector is only sensitive to ions, the photofragments need to be ionized without changing their directions by irradiating another intense femtosecond probe pulse after the dissociation is complete. Some typical ion images of I^+ fragments recorded for I_2 molecules are shown in Fig. 1.

They have shown that the degree of alignment is increased by lowering the initial rotational temperature of the molecules or by increasing the intensity of the alignment laser field, which is consistent with theoretical expectations. They have further confirmed that the degree of alignment follows the intensity profile of the alignment laser pulse. The same group has demonstrated the molecular alignment of other molecules such as ICl, CS_2, CH_3I, and C_6H_5I. They have also discussed the potential applications of a sample of aligned molecules, which include dissociative

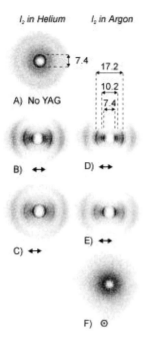

Fig. 1. Typical ion images of I^+ fragments recorded for I_2 molecules seeded in He (left images) and Ar (right images), and for different intensities of the YAG pulse used as the alignment pulse. (A) $I_{YAG} = 0$. (B), (D), and (F) $I_{YAG} = 1.4 \times 10^{12}$ W/cm^2. (C) and (E) $I_{YAG} = 5.0 \times 10^{11}$ W/cm^2. The polarization of the femtosecond probe pulse is parallel to the molecular beam axis, i.e., perpendicular to the page and the polarization of the YAG pulse is shown at each image. The length scales in (A) and (D) are given in millimeters. In all images, the central portion (radius ≤2.5 mm) of the detector is inactive. (Reprinted with permission from Sakai et al.[13] Copyright (1999) by the American Institute of Physics.)

rotational cooling, control of strong laser field ionization, selective dissociation, applications in chemical control, orientational effects in bimolecular reactions, and alignment of the molecular plane.[15] Among them, Larsen et al. have demonstrated selective dissociation[16] and alignment of the molecular plane (3D alignment)[17] as described below.

As an interesting extension of the molecular alignment, Karczmarek et al. proposed an optical centrifuge with which all anisotropic molecules can be spun and even dissociated by an intense rotating linearly polarized laser field.[18] They further suggest that the optical centrifuge can be used to efficiently separate isotopes because the centrifugal effect is sensitive to the moment of inertia and the heavier species will dissociate first. Villeneuve et al. demonstrated the forced rotation of Cl_2 molecules in the optical centrifuge.[19] A linearly polarized field rotating at frequency Ω is shaped by combining two counterrotating circularly polarized laser fields with frequencies $\omega_L - \Omega$ and $\omega_L + \Omega$. In order to increase Ω with time, they chirp the two circularly polarized pulses in opposite frequency directions. They thus accelerate the rate of polarization rotation from 0 to 6 THz in 50 ps, spinning Cl_2 molecules from near rest up to the states with angular momentum $J \sim 420$. At the highest rotation rate, the molecular bond is broken and the molecule dissociates. Concerning the molecular alignment techniques in the adiabatic regime, the readers should refer to Ref. 20.

In the case of adiabatic molecular alignment, a sample of aligned molecules can be maintained for a relatively long period of typically a few nanoseconds in the strong laser field. Alternatively, theoretical calculations have revealed that an intense short laser pulse can be used to excite rotational wavepackets through impulsive Raman processes and induce molecular alignment.[21] After the pump pulse, the molecular alignment is dephased and revived at the distinct time delays after the pump pulse, which are characteristic of the specific molecules. Rosca-Pruna and Vrakking excited rotational wavepackets in I_2 molecules with an intense picosecond laser pulse and observed revival structures after the pump pulse with a velocity-map ion imaging technique.[22] With respect to the molecular alignment techniques in the nonadiabatic regime, the readers should refer to Ref. 23.

One of the intriguing subjects of nonadiabatic molecular alignment is to enhance the degree of alignment by shaping the ultrashort pump pulse. Suzuki *et al.* recently demonstrated the optimal control of nonadiabatic alignment of N_2 molecules by shaping the femtosecond pump pulses with the direct feedback of the degree of alignment, which is evaluated by the 2D ion imaging technique.[24] They find that the alignment is optimized by doubly peaked pulses with approximately equal intensities. Based on the considerations from both time and frequency domains, they conclude that a doubly peaked pulse with an appropriate interval can be regarded as a single pulse with a center trough and that the effective duration of a doubly peaked pulse rather than its structure is of crucial importance to optimize nonadiabatic molecular orientation.

2.2 1D and 3D alignments

The spatial directions of a molecule can be determined by a set of three Euler angles (θ, ϕ, and χ). The interaction between a linearly polarized laser field along the space-fixed z-axis and the induced dipole moment leads to 1D alignment where only one Euler angle θ is confined around the z-axis. When planar molecules are used as a sample, 3D molecular alignment, *i.e.*, the control of all three Euler angles is desirable and can be achieved by using an elliptically polarized laser field, which was first suggested in Ref. 15. Since an elliptically polarized field can be considered to consist of two perpendicular and linearly polarized fields with a large and small amplitude, a planar molecule can be expected to be confined to the elliptical polarization plane with the molecular axis with the largest polarizability component parallel to the major axis of the elliptical polarization and with the second largest polarizability component parallel to the minor axis of the elliptical polarization, when the field-molecule interaction potential is minimized both along the major axis and along the minor axis of elliptical polarization. Along this approach, Larsen *et al.* have demonstrated 3D alignment of 3,4-dibromothiophene molecules with an intense, nonresonant, and elliptically polarized laser field.[17]

For polyatomic asymmetric top molecules, three different rotational constants lead to three different rotational periods. Therefore, we cannot expect the realization of field-free 3D molecular alignment *via* the rotational revivals. Without relying on the use of rotational revivals, Lee *et al.* demonstrated field-free 3D alignment of SO_2 molecules by using two time-delayed, orthogonally polarized, nonresonant, and femtosecond laser pulses.[25] The molecular alignment is observed with the time-delayed coincidence Coulomb explosion imaging.

3. Techniques for Molecular Orientation

3.1 *1D and 3D orientations*

In general, the control of molecular orientation has been thought to be much more difficult than that of molecular alignment. In order to achieve molecular orientation, the field–molecule interaction must have an asymmetric potential, which can be created by adding an electrostatic field to an intense and nonresonant laser field. In fact, Friedrich and Herschbach suggested that molecular orientation in a weak electrostatic field can be greatly increased by combining an intense and nonresonant laser field.[26,27] Along this combined-field approach, Sakai and his coworkers have demonstrated 1D orientation of OCS molecules ($\mu = 0.71$ D) with combined electrostatic and intense, nanosecond, and linearly polarized laser fields.[28,29] As a measure of molecular orientation, they observe asymmetric ion yields of forward and backward fragments produced by the Coulomb explosion with intense femtosecond probe pulses polarized parallel to the time-of-flight axis. It is confirmed that the degree of orientation can be increased by increasing the electrostatic field and the intensity of the laser field or by decreasing the initial rotational temperature of the molecules. Furthermore, they have demonstrated 3D orientation of 3,4-dibromothiophene molecules with combined electrostatic and intense, nanosecond, and elliptically polarized laser fields.[30] 3D molecular orientation is confirmed by the complementary observations of both 3D alignment with 2D ion imaging and 1D orientation with time-of-flight mass spectrometry.

3.2 *Laser-field-free molecular orientation*

The above-mentioned demonstrations of 1D and 3D molecular orientations have been achieved in the adiabatic regime. Since an intense laser field can modify the physics and/or chemistry involved, the realization of nonadiabatic laser-field-free molecular orientation had been desired for a long time. Although most of the theoretical proposals to this end are based on utilizing an intense half-cycle pulse in the terahertz region as a quasi-electrostatic field,[31,32] laser-field-free molecular orientation along those schemes has remained unrealized. This may be due to the difficulties of both generation and control of a required intense half-cycle pulse of $\sim$150 kV/cm or more. In addition, a small but long field component with opposite polarity, which inevitably follows the main half-cycle pulse, can be fatal in those schemes.

Alternatively, Sakai and his coworkers suggested that laser-field-free molecular orientation can be achieved with the combination of an electrostatic field and an intense and nonresonant laser field with a rapid turnoff.[33] The orientation achieved at the peak of the laser pulse is expected to revive at the rotational period of the molecule with the same degree of orientation. In this strategy, the sufficiently long rising time of the laser pulse is of crucial importance to ensure purely adiabatic molecular orientation and achieve the highest possible degree of orientation. In fact, when significant alignment is achieved by a deficiently long pulse for adiabatic orientation, a significantly deep but insufficiently asymmetric double-well potential is created and prevents rotational quantum states in molecules from efficiently tunneling from a shallower potential well to a deeper one. Along their own strategy, they have recently demonstrated laser-field-free orientation of OCS molecules with the combination of a weak electrostatic field and an intense and nonresonant rapidly turned-off laser field,[34] which can be shaped with the plasma shutter technique.[35] Molecular orientation is adiabatically created in the rising part of the laser pulse, and it is found to revive at around the rotational period of an OCS molecule with the same degree of orientation as that at the peak of the laser pulse. The temporal evolutions of molecular orientation at around the turnoff of the pump pulse and the rotational period of an OCS molecule are shown in Fig. 2, where a parameter FB $\equiv (I_\mathrm{f} - I_\mathrm{b})/(I_\mathrm{f} + I_\mathrm{b})$ with I_f and I_b the integrated signals of the forward and the backward fragments, respectively is used as a measure of the degree of orientation. In the case of OCS, the FB parameter for S^{3+} ions and the orientation cosine $\langle\cos\theta\rangle$ behave in the same way.

3.3 All-optical molecular orientation

In the above demonstration of laser-field-free molecular orientation, a weak electrostatic field exists after the rapid turnoff of the laser pulse though it does not play any role in reviving molecular orientation at around the rotational period of the molecule and is negligible compared to the laser electric field. This is due to the fact that we have no technique to turnoff an electrostatic field faster than $\sim$200 fs. In order

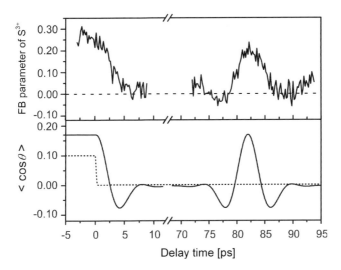

Fig. 2. The upper panel shows the temporal evolutions of the FB parameter for S^{3+} fragment ions at around the turnoff of the pump pulse and the rotational period ($T_{\rm rot} = 82$ ps) of an OCS molecule. The lower panel shows the calculations of the degree of orientation $\langle\cos\theta\rangle$, where θ is the polar angle between the electrostatic field and the permanent dipole moment of an OCS molecule. The intensity profile of the shaped pump pulse is also represented by a dotted curve in the lower panel. (Reprinted with permission from Goban et al.[34] Copyright (2008) by the American Physical Society.)

to achieve completely field-free molecular orientation, we need a technique that does not utilize an electrostatic field. Kanai and Sakai have proposed an approach to achieve molecular orientation with an intense, nonresonant, and two-color laser field in the adiabatic regime,[36] which relies on the combined effects of anisotropic hyperpolarizability interaction as well as anisotropic polarizability interaction. A two-color and linearly polarized laser electric field is expressed as

$$E(t) = E_\omega(t)\cos\omega t + E_{2\omega}(t)\cos(2\omega t + \phi), \quad (1)$$

with $E_\omega(t)$ and $E_{2\omega}(t)$ the electric field amplitude profiles of the ω and 2ω fields, respectively, and ϕ the phase difference between the two wavelengths. After the cycle average of the rapid oscillation of the laser electric field, the total Hamiltonian of a linear molecule interacting with the asymmetric field [Eq. (1)] is given by

$$H(t) = B\mathbf{J}^2 - \frac{1}{4}[\alpha_{zz}\cos^2\theta + \alpha_{xx}\sin^2\theta][E_\omega^2(t) + E_{2\omega}^2(t)]$$
$$- \frac{1}{8}[\beta_{zzz}\cos^3\theta + 3\beta_{zxx}\cos\theta\sin^2\theta]\cos\phi E_\omega^2(t)E_{2\omega}(t), \quad (2)$$

with B the rotational constant of the molecule, $\mathbf{J}^2$ the squared angular momentum operator, θ the angle between the laser polarization and the molecular axis, α_{zz} and α_{xx} the polarizability components parallel and perpendicular to the molecular axis, respectively, and β_{zzz} and β_{zxx} the hyperpolarizability components parallel and perpendicular to the molecular axis, respectively. The mechanism of all-optical molecular orientation can be understood by referring to the Hamiltonian [Eq. (2)]. The double-well potential is created mainly by the anisotropic polarizability interaction given by the second term in Eq. (2) and the asymmetry in the potential, which is essential for molecular orientation, is brought by the anisotropic hyperpolarizability interaction given by the third term in Eq. (2). It should be noted that the permanent dipole interaction vanishes in Eq. (2) after the cycle average.

Recently, Sakai and his coworkers have demonstrated clear evidence of all-optical orientation of OCS molecules with an intense nonresonant two-color laser field in the adiabatic regime.[37] A two-color and linearly polarized laser electric field is prepared by the superposition of the fundamental pulse (wavelength $\lambda_\omega = 1064$ nm) from an injection-seeded nanosecond Nd:YAG laser (pulse width $\tau \sim 12$ ns) and its second harmonic pulse ($\lambda_{2\omega} = 532$ nm). The relative phase between the two wavelengths is controlled by the rotation of a 6.35-mm-thick fused silica plate. The two wavelengths are collinearly focused into the vacuum chamber by a 250-mm-focal-length plano-convex achromatic lens. The typical intensity is $I \sim 1 \times 10^{12}$ W/cm^2 for each wavelength. The sample is OCS molecules 5% diluted with Ar, which is supplied as a supersonic molecular beam by a pulsed valve equipped with a 0.25-mm-diameter nozzle. Since the technique does not need the permanent dipole interaction for molecular orientation, the polarizations of the two-color pump laser fields are set parallel to the detector plane. Thereby the molecular orientation is observed with the 2D ion-imaging technique. An intense femtosecond probe pulse ($\lambda \sim 800$ nm, $\tau \sim 100$ fs, $I \sim 3 \times 10^{14}$ W/cm^2) is collinearly focused into the vacuum chamber with the two-color pump pulse and is used to multiply ionize the OCS molecules at the peak of the two-color pump pulse. The polarization of the probe pulse is set perpendicular to the detector plane to avoid the alignment dependence in the multiphoton ionization process.

When the relative phase is zero, the upper part of CO$^+$ image is more intense than the lower part, and the lower part of S$^+$ image is more intense than the upper part. In contrast, when the relative phase is pi, the lower part of CO$^+$ image is more intense than the upper part, and the upper part of S$^+$ image is more intense than the lower part. These observations clearly show that the molecular orientation can be changed simply by changing the relative phase between the two wavelengths. They further observe the degree of molecular orientation $\langle \cos \theta_{2D} \rangle$ with θ_{2D}, the projection of the polar angle between the polarization of the two-color pump pulse and the molecular axis onto the detector plane as a function of the relative phase between the two wavelengths. The results are shown in Fig. 3, where $\langle \cos \theta_{2D} \rangle$'s for CO$^+$ and S$^+$ are represented by the red circles and black squares, respectively. When $\langle \cos \theta_{2D} \rangle$ is not saturated and proportional to the potential difference between the deeper and shallower wells, $\langle \cos \theta_{2D} \rangle$ is expected to follow $\cos \phi$ according to

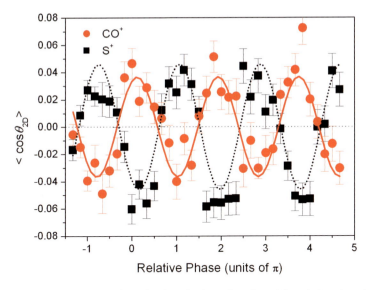

Fig. 3. The degree of molecular orientation $\langle\cos\theta_{2D}\rangle$ as a function of the relative phase between the two-color laser fields. The $\langle\cos\theta_{2D}\rangle$'s for CO^+ and S^+ are represented by the red circles and black squares, respectively. The red solid and black dotted curves are the least squares fits of the observations. (Reprinted with permission from Oda et al.[37] Copyright (2010) by the American Physical Society.)

Eq. (2). The red solid and black dotted curves in Fig. 3 are the least squares fits of the observations. One can see that $\langle\cos\theta_{2D}\rangle$ for CO^+ modulates out of phase with that for S^+ and that molecular orientation can be changed every pi change in the relative phase between the two wavelengths. These observations provide conclusive evidence of molecular orientation with an intense nonresonant two-color laser field. They have also observed orientation of C_6H_5I molecules with the present technique, which serves to prove the fact that the technique is versatile and applicable to various molecules because it utilizes the nonresonant interactions.

By applying the technique with a switched pulse to the all-optical two-color method, Muramatsu et al. proposed a practical and versatile technique to achieve *completely field-free* molecular orientation with an intense, nonresonant, and two-color laser field with a slow turn on and rapid turnoff.[38] On the other hand, De et al. claim that they have observed nonadiabatic field-free orientation of CO molecules by an intense two-color (800 and 400 nm) femtosecond laser field.[39] It should be noted that higher pump pulse intensities and the probe pulse polarization parallel to the pump pulse polarization, which should have led to the significant ionization of CO molecules and the significant enhancement of the apparent degrees of molecular orientation due to the angle-dependent multiphoton ionization and dynamic realignment of CO molecules, respectively, are used in their observations. It has been shown theoretically that when the thermal ensemble is used as a sample the degrees of molecular orientation achieved with an intense two-color femtosecond laser field are very low.[38] It should be extremely difficult to measure these very low degrees of molecular orientation with the Coulomb explosion imaging. Therefore, it is absolutely necessary to reverify whether the results suggesting nonadiabatic field-free molecular orientation survive by lowering the pump pulse intensities and setting the probe pulse polarization perpendicular to the detector plane.

4. Various Applications with a Sample of Aligned or Oriented Molecules

In this section, we review various applications with a sample of aligned or oriented molecules to prove its usefulness.

4.1 *Tunnel ionization*

Tong *et al.* developed a theoretical model of molecular tunneling ionization, which can predict alignment dependence of tunneling ionization rates in molecules.[40] The so-called Ammosov–Delone–Krainov (ADK) model developed for atoms[41] is extended to a model for diatomic molecules by taking account of the symmetry property and the asymptotic behavior of the molecular electronic wavefunction. Their molecular ADK (MO-ADK) theory is applied to calculate the ratios of ionization yields for several diatomic molecules to those of their companion atoms that have almost the same ionization potential as that for the given valence orbital. They show that the predicted ratios for both pairs with ionization suppression (D_2/Ar, O_2/Xe) and pairs without ionization suppression (N_2/Ar, CO/Kr) are in good agreement with the experimental results. They also show that the MO-ADK model works well for NO, S_2, and SO molecules with π orbitals. For D_2, the smaller electronic charge density in the asymptotic region is responsible for the ionization suppression, while for molecules such as O_2 the ionization suppression originates in the fact that the valence electrons are in the π orbitals. In fact, since the electronic cloud for the latter molecules is perpendicular to the molecular axis, the tunnel ionization rate is small when the molecules are aligned along the laser field direction. Although the MO-ADK model predicts the ionization suppression for the pair of F_2 (with π orbital)/Ar, the experimental results show no suppression, which is the only exception and still an open question.

Litvinyuk *et al.* experimentally investigated alignment dependence in strong field single ionization of neutral N_2 molecules.[42] A 40-fs linearly polarized pump pulse with the intensity of 5×10^{13} W/cm^2 is used to create rotational wavepackets in molecules and induce nonadiabatic molecular alignment. By Coulomb exploding the molecules with an intense circularly polarized probe pulse (4×10^{15} W/cm^2) and observing the N^{3+} fragment ions produced from the asymmetric N$_2^{5+} \rightarrow$ N^{3+} + N^{2+} dissociation channel, they first measure temporal evolution of the molecular alignment parameter $\langle \cos^2 \vartheta \rangle$ for N$_2$ molecules, where ϑ is the angle between the polarization direction of the pump pulse and the projection of the molecular axis onto the circular polarization plane of the probe pulse (in order to understand their experimental setup clearly, the readers should refer to Fig. 3 in Ref. 43). Having known the temporal dependence of molecular alignment distributions, another probe pulse with the intensity of 2×10^{14} W/cm^2 is used to measure the total N$_2^+$ ion yield at two pump-probe delays $\tau = 4.15$ ps and $\tau = 4.33$ ps, when the molecules are best aligned parallel and perpendicular to the field direction of the probe pulse. They find that the ionization probability for N$_2$ molecules aligned parallel to the laser electric field is about four times larger than for those aligned perpendicular to the laser electric field. They also show that theoretical calculations by Tong and Lin[44] are in good agreement with their experimental results.

4.2 Nonsequential double ionization

Zeidler et al. investigated the dynamics of nonsequential double ionization in nonadiabatically aligned N$_2$ molecules.[3] They create rotational wavepackets in N$_2$ molecules with a 60-fs pump pulse ($\sim 2.5 \times 10^{13}$ W/cm^2). The molecules are ionized by a delayed intense 40-fs probe pulse ($\sim 1.2 \times 10^{14}$ W/cm^2) at two pump-probe delays when N$_2$ molecules are aligned (~ 3.93 ps) and anti-aligned (~ 4.30 ps). The polarization direction of the ionization pulse is parallel to the spectrometer axis and perpendicular to the polarization direction of the alignment pulse. Therefore, the alignment distribution is perpendicular at ~ 3.93 ps and parallel at ~ 4.30 ps to the polarization of the ionization pulse. They employ cold target recoil ion momentum spectroscopy (COLTRIMS)[45] to determine the momenta of ions and electrons. They first confirm that both electrons are most likely to be emitted with a similar momentum $k_a^{\parallel} = k_b^{\parallel} \approx 0.5$ a.u. in the same direction for both alignment distributions, where $k_a^{\parallel}$ and $k_b^{\parallel}$ are the momentum component parallel to the laser polarization direction of one electron and that of the other electron, respectively. They also show that double ionization is slower for perpendicular molecules. More specifically, both electrons are more likely to be ejected from the constituent atoms rapidly within the same half laser cycle for parallel molecules in the same direction, while the emission of one of the two electrons is more likely to be delayed for perpendicular molecules, and the electrons tend to be ejected during different laser half cycles. That is, when the electrons are well defined in a potential with one degree of freedom along the laser polarization direction, they assist each other in leaving parallel molecules. This mechanism is called Coulomb-assisted ionization.

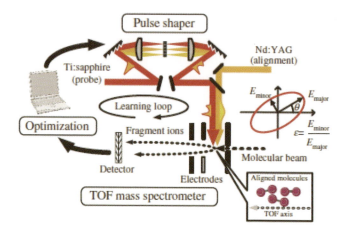

Fig. 4. An experimental setup to control multiphoton ionization processes in aligned I_2 molecules with time-dependent polarization pulses. A pulse shaper is used to shape femtosecond pulses for learning-loop optimal quantum control. Experimental feedback signals are collected from time-of-flight mass spectra that result from the interaction of the shaped femtosecond pulses with aligned molecules. These signals are processed in a computer to iteratively improve the applied pulse shape based on genetic algorithm until an optimal result is obtained. The definitions of ellipticity ε and the orientation angle θ are also illustrated. (Reprinted with permission from Suzuki et al.[46] Copyright (2004) by the American Physical Society.)

4.3 Control of multiphoton ionization

With a sample of *adiabatically* aligned I_2 molecules, Suzuki et al. demonstrated optimal control of multiphoton ionization processes in I_2 molecules.[46] In their demonstration, a time-dependent polarization pulse, where a polarization state in a femtosecond pulse changes as a function of time, is introduced as a novel control parameter.[47] Figure 4 shows an experimental setup to control multiphoton ionization processes in aligned I_2 molecules with time-dependent polarization pulses. A time-dependent polarization pulse enables us to take full advantage of vectorial nature of a laser electric field. They find a correlation between a femtosecond time-dependent polarization pulse and the production efficiency of evenly or oddly charged molecular ions. Much better controllability of the correlation is achieved with a time-dependent polarization pulse than with a pulse having a fixed ellipticity, which suggests the existence of an uninvestigated tunnel ionization mechanism that is characteristic of a time-dependent polarization pulse.[48] Their study points to some new directions in optimal control studies on molecular systems: (1) A sample of aligned molecules is employed in the optimal control experiment for the first time. (2) In order to optimize a quantum process in molecules, a time-dependent polarization pulse is introduced to the learning-loop optimal control system for the first time. (3) By using a sample of aligned molecules and a time-dependent polarization pulse, both external and internal degrees of freedom in molecules are simultaneously controlled.

4.4 Control of photodissociation

Using a sample of *adiabatically* aligned I_2 molecules, Larsen et al. controlled the branching ratio between the $I+I$ and $I+I^*$ photodissociation channels up to a factor of 26.[16] The I_2 molecules are adiabatically aligned by the irradiation of a 3.5-ns-long pulse from a pulsed single-longitudinal mode Nd:YAG laser ($\lambda = 1064$ nm). When the aligned I_2 molecules are irradiated by a 150-fs-long pump pulse centered at 485 nm, the molecules are either excited to the $B\,^3\Pi\,0_u^+$ state that dissociates into $I+I^*$ or to the $^1\Pi\,1_u^+$ state that dissociates into $I+I$, where I and I^* stand for iodine atoms in the $^2P_{3/2}$ and $^2P_{1/2}$ states, respectively. Since the $B\,^3\Pi\,0_u^+$ ($^1\Pi\,1_u^+$) state is connected to the $X^1\Sigma\,0_g^+$ ground state with a parallel (perpendicular) transition moment, the branching ratio between the two dissociation channels defined as $R = (I+I)/(I+I^*)$ can be controlled by changing the angle α between the YAG polarization and the polarization of the pump pulse. They measure the yield of the $I+I$ and the $I+I^*$ dissociation channels by changing α from $0°$ to $90°$ as shown in Fig. 5 and find that the branching ratio R reaches its extreme values at $\alpha = 0°$ and $90°$: $R(\alpha = 0°) = 0.27 \pm 0.02$ and $R(\alpha = 90°) = 6.9 \pm 0.5$, which means that

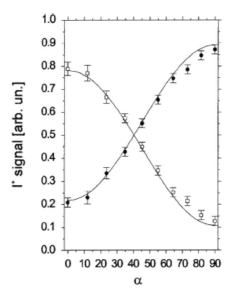

Fig. 5. The yields of the $I+I$ (solid circles) and the $I+I^*$ (open squares) dissociation channels as a function of the angle α between the polarizations of the YAG pulse and the pump pulse. The solid curves represent the calculated yields of the two dissociation channels. (Reprinted with permission from Larsen et al.[16] Copyright (1999) by the American Physical Society.)

the contrast is 26. In this demonstration, it is essentially important that the excited states correlate adiabatically with the final product states.

4.5 High-order harmonic generation

Itatani *et al.* demonstrated that the full 3D structure of a single-electron wavefunction of an N_2 molecule, *i.e.*, a highest occupied molecular orbital (HOMO) can be imaged by observing high-order harmonics from aligned molecules with the irradiation of intense femtosecond laser pulses.[4] Such tomographic imaging of a molecular orbital relies on three important steps. (1) The molecular axis is nonadiabatically aligned in the laboratory frame. (2) The orbital with the lowest ionization potential, *i.e.*, the HOMO of a molecule in the ground state is preferentially ionized because the tunnel ionization rate depends exponentially on the ionization potential. (3) A series of projected images of the molecular orbital can be taken by observing high-order harmonics at a series of molecular alignments. What they observe is the transition dipole moment from the HOMO to a coherent set of continuum wavefunctions. The procedures of tomographic reconstruction of the molecular orbital can be summarized as follows: The angular frequency ω of the harmonic field and the wavenumber k of the electron de Broglie wave are connected by $k(\omega) = (2\omega)^{1/2}$ in atomic units. The transition dipole moment between the orbital wavefunction and the electron wavefunction in the continuum approximated by the plane wave is given by $\mathbf{d}(\omega; \theta) = \langle \Psi(\mathbf{r}; \theta)|\mathbf{r}|\exp[ik(\omega)x]\rangle$ with $\Psi(\mathbf{r}; \theta)$ the orbital wavefunction rotated by the Euler angle θ. Under the condition of the perfect phase matching, the harmonic signal is given by $S(\omega; \theta) = N^2(\theta)\omega^4|a[k(\omega)]\mathbf{d}(\omega; \theta)|^2$, where $a[k(\omega)]$ is the complex amplitude of component k of the continuum wavepacket and $N(\theta)$ is the number of the produced ions. The value of $a[k(\omega)]$ is experimentally obtained by measuring $S_{\text{ref}}(\omega)$ for a reference Ar atom with $2p_x$ orbital because $d_{\text{ref}}(\omega) = \langle \Psi_{\text{ref}}(r)|r|\exp[ik(\omega)x]\rangle$ is known, and given by $a[k(\omega)] = \omega^{-2}[S_{\text{ref}}(\omega)]^{1/2}[|\langle \Psi_{\text{ref}}(r)|r|\exp[ik(\omega)x]\rangle|^2]^{-1/2}$. Then, $S(\omega)$ is measured for a target molecule, giving its transition dipole moment $d(\omega; \theta)$. Since the definition of $d(\omega; \theta)$ can be regarded as a spatial Fourier transform of the orbital along the x direction, the wavefunction Ψ is obtained by doing the inversion transform with both the Fourier slice theorem[49] and one assumption that the factor $\mathbf{r}$ included in the Fourier transform and defined in the laboratory frame does not rotate with the molecular frame θ. One must also consider the angular dependence of the ionization rate to normalize each harmonic spectrum. The angular dependence of the ionization rate can be known either by the MO-ADK theory or by measuring the angular dependence of the ion yield $N(\theta)$. The reconstructed molecular orbital wavefunction of an N_2 molecule is shown in Fig. 6(a), which can be compared with the *ab initio* orbital calculation of the N_2 $2p\sigma_g$ orbital shown in Fig. 6(b).

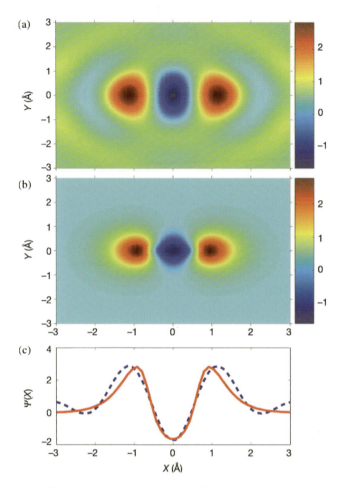

Fig. 6. Molecular orbital wavefunctions of an N_2 molecule. (a) The reconstructed wavefunction of the HOMO of an N_2 molecule. The reconstruction is obtained by a tomographic inversion of the high-order harmonic spectra observed at 19 projection angles. The fact that both positive and negative values are present shows that this is not the square of a wavefunction but a wavefunction up to an arbitrary phase. (b) The shape of the N_2 $2p\,\sigma_g$ orbital wavefunction from an *ab initio* calculation. The gray scales are the same for both images. (c) The cuts along the internuclear axis for the reconstructed (dashed) and *ab initio* (solid) wavefunctions. (Reprinted with permission from Itatani *et al.*[4] Copyright (2004) by the Nature Publishing Group.)

Kanai *et al.* demonstrated quantum interference in the recombination process with a sample of nonadiabatically aligned CO_2 molecules.[5] They perform a pump-probe experiment, where a pump pulse is used to create rotational wavepackets and induce nonadiabatic molecular alignment and a probe pulse is used to generate high-order harmonics. They observe both the harmonic intensity and the ion yield under the same experimental condition as a function of the pump-probe delay, which serve to disentangle the contributions from the ionization and recombination processes.

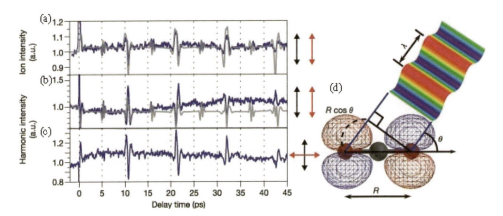

Fig. 7. The temporal evolution of the ion yield and the 23rd harmonic intensity from CO_2 molecules, and an illustration of the model of two point emitters. (a–c) The ion yield dominated by CO_2^+ (a) and the 23rd harmonic intensities from CO_2 molecules (b, c) as a function of the pump-probe delay. The polarizations of the pump and probe pulses are parallel (a, b) or perpendicular (c) to each other. The rotational period T_{rot} of a CO_2 molecule is 42.7 ps. The results of theoretical calculations are also shown by gray curves. The rotational temperature of the CO_2 molecules is assumed to be 40 K and the highest degree of alignment $\langle \cos^2 \theta \rangle$ is estimated to be 0.70 at $3T_{rot}/4$. (d) A CO_2 molecule can be regarded as an elongated diatomic molecule. Two point emitters are located in two O nuclei. λ is the de Broglie wavelength of an electron. θ is the orientation angle, *i.e.*, the angle between the molecular axis and the polarization of the probe pulse. R and $R \cos \theta$ are the distance between two O atoms and its projection, respectively. (Reprinted with permission from Kanai et al.[5] Copyright (2005) by the Nature Publishing Group.)

For N_2 and O_2 molecules, the harmonic intensity is found to modulate in phase with the ion yield with the polarizations parallel to each other. For CO_2 molecules, however, they find that the harmonic intensity modulates out of phase with the ion yield with the polarizations parallel to each other as shown in Fig. 7. By applying the model of two point emitters, which was originally proposed for a diatomic molecule, to a triatomic molecule, they successfully ascribe the observed anticorrelation between the harmonic intensity and the ion yield to quantum interference in the recombination process.

Based on the nature of quantum interference during high-order harmonic generation from aligned molecules, they propose a novel approach to probe the instantaneous structure of a molecular system.[5,50] The procedures of this new probe can be summarized as follows: (1) Prepare a sample of well-aligned molecules. (2) Observe the angular dependence of the ion yield ($I_{ion}(\theta)$ with θ being the angle between the molecular axis and the probe pulse polarization) and the harmonic intensity ($I_{HH}(\theta)$) of an appropriate harmonic order, for which destructive interference can be expected to take place, from aligned molecules by rotating the probe pulse polarization. (3) Plot a graph of the ratio $I_{HH}(\theta)/I_{ion}(\theta)$ against θ, which should give the angular dependence of the recombination process $I_{recom}(\theta)$ because the $I_{HH}(\theta)$ is proportional to the product of $I_{ion}(\theta)$ and $I_{recom}(\theta)$. (4) Find the angle θ at which a dip appears in the graph. (5) The instantaneous bond length R can then be determined by the condition for destructive interference, i.e., $R\cos\theta = \lambda$ when the molecule's HOMO has anti-bonding symmetry, where λ is the de Broglie wavelength of a free electron.

Kanai et al. also observed two distinct effects on ellipticity dependence of high-order harmonic generation from aligned N_2, O_2, and CO_2 molecules: (1) The ellipticity dependence is sensitive to molecular alignment and to the shape and symmetry of the valence orbitals. (2) The destructive interference in the recombination process also affects the ellipticity dependence.[6] With respect to the ultrafast molecular imaging based on high-order harmonic generation from (aligned) molecules, the readers should refer to Ref. 51.

4.6 Structural deformation of polyatomic molecules

With a sample of nonadiabatically aligned CO_2 molecules, Minemoto et al. revealed alignment dependence of structural deformation in the production process of multiply charged CO_2 molecular ions in an intense femtosecond laser field.[52] They first confirm that the structures of both neutral CO_2 in the evolution of the rotational wavepacket and CO_2^+ produced in the course of the production of CO_2^{2+} are linear as in the ground state of neutral CO_2 molecules, which is ensured by the satisfactory agreement between the observed temporal evolution of the degree of alignment of CO_2 molecules as a function of the pump-probe delay Δt and the theoretical simulation. The best aligned CO_2 molecules at around half revival of $\Delta t = 20.9$ ps are used as sample molecules. By using the 2D ion imaging technique, they observe the C^+ fragment ions produced from the channel of $O^+ + C^+ + O^+$ dissociated from CO_2^{3+} and emitted $90° \pm 10°$ and $270° \pm 10°$ to the pump pulse polarization for the evaluation of the bending angle σ defined in Fig. 8(a). Thus, they measure the distributions of the momentum component parallel to the detector plane of C^+ ions emitted within the above angles for different angles θ between the pump pulse polarization and the probe pulse polarization. The maximum momentum of C^+ ions can be transformed to the maximum bending angle σ_{max} by using the law of conservation of momentum. As shown in Fig. 8(b), they find that σ_{max} decreases monotonically from $24°$ for $\theta = 0°$ to $16°$ for $\theta = 90°$. The observed alignment dependence of structural deformation is explained by the field-induced nonadiabatic transition between the lowest two adiabatic states of CO_2^{2+} ions.[53]

4.7 Selective preparation of one of the enantiomers

Fujimura et al. proposed a novel approach to achieve selective preparation of one of enantiomers with specific chirality from a 50:50% racemate consisting of two

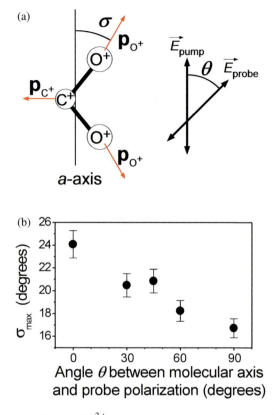

Fig. 8. (a) An illustration of a bent CO_2^{3+} molecular ion just before Coulombic dissociation and the definitions of σ and the a-axis together with the angle θ between the pump and the probe polarizations. (b) Alignment dependence of the maximum bending angle σ_{max} as a function of the angle θ between the molecular axis and the polarization of the probe pulse. The σ_{max} decreases monotonically from the parallel configuration to the perpendicular configuration. (Reprinted with permission from Minemoto et al.[52] Copyright (2008) by the American Physical Society.)

equivalent enantiomers with opposite chiralities by the irradiation of an optimally designed time-dependent polarization pulse.[54] In fact, the selective preparation of one of enantiomers with specific chirality is one of intriguing subjects both in chemistry and biochemistry. Their sample molecule H_2POSH has been chosen based on the four criteria: (i) small size and simple structure, (ii) sufficiently high potential energy barrier separating the two enantiomers, (iii) substantial variation of the relevant molecular dipole components along the path from one enantiomer configuration to the other, and (iv) chemical interest. By means of the method of local optimal control, they demonstrate that the molecular wavepacket in the initial state corresponding to the 50:50% racemate can be driven to the nearly pure single enantiomer by the optimally designed time-dependent polarization pulse, which consists of four sub-pulses. The time-dependent population dynamics reveals the underlying two processes for the present laser-induced preparation of pure enantiomers. The first process from the first three sub-pulses is to create an appropriate superposition of two excited states from the torsional ground state. The second process from the last sub-pulse is to transfer this superposition of the excited states to the appropriate superposition of delocalized states with the same or opposite phases, thus leading to the preparation of the localized target states. In order to implement the proposed scenario, there are two experimental prerequisites: (1) the preparation of a sample of oriented molecules and (ii) the shaping of an optimally designed time-dependent polarization pulse, both of which are now within the reach of the state-of-the-art experimental techniques.

4.8 Phase effects and attosecond science

Lan et al. has demonstrated that a sample of oriented molecules can be a unique nonlinear medium to generate only one attosecond pulse in each cycle of the driving laser field with the same carrier-envelope phase from pulse to pulse.[55] Unlike in atoms and symmetric molecules, in asymmetric molecules, an electron is preferentially localized in the deeper well and the ionization process (the first step in the three-step model for high-order harmonic generation) is quite different in consecutive half cycles of the driving laser field (Fig. 9). When the laser electric field is parallel to the permanent dipole moment of the molecule, the electron energy is decreased in the potential modified by the laser field. Then, the ionization probability becomes low and consequently harmonic generation is suppressed. In the consecutive half cycle, however, the laser field changes its sign and is antiparallel to the permanent dipole moment of the molecule, when the electron energy is increased in the potential modified by the laser field. Then, the ionization probability is enhanced and the harmonic generation is also enhanced. Since the harmonic and attosecond pulse yields in the latter antiparallel case are much higher than those in the former parallel case, only one attosecond pulse can be generated in a full cycle of the driving laser field with the same carrier-envelope phase from pulse to pulse. This scenario has been verified by assuming that the molecule is aligned along the laser polarization direction (1D model) and by numerically solving the time-dependent Schrödinger equation. It should be noticed that one has to prepare a sample of *oriented* molecules to implement the present scheme.

4.9 Polarizability anisotropies of rare gas van der Waals dimers

Minemoto et al. applied the laser-induced molecular alignment technique in the adiabatic regime to the evaluation of polarizability anisotropies of rare gas van der

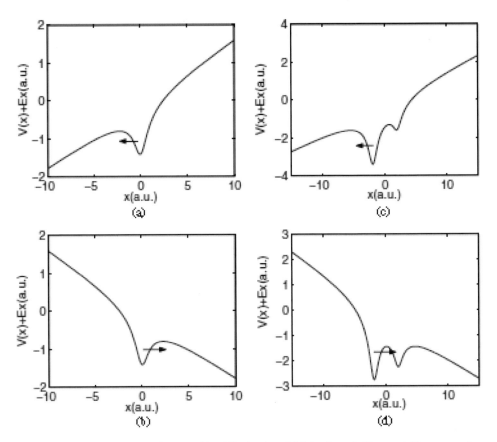

Fig. 9. Schematic diagrams of the combined Coulomb and driving laser field potential in consecutive half driving laser cycles for atoms (a) and (b) and asymmetric molecules (c) and (d). (Reprinted with permission from Lan et al.[55] Copyright (2007) by the American Physical Society.)

Waals dimers Rg_2 ($Rg = Ar$, Kr, and Xe).[56] They first confirm that the weakly bound rare gas dimers Rg_2 can be aligned by an intense nonresonant laser field and find that the degree of alignment $\langle \cos^2 \theta \rangle$, where θ is the angle between the pump pulse polarization and the molecular axis, becomes larger in order of Ar_2, Kr_2, and Xe_2, which reflects the order of magnitudes of their polarizability anisotropy $\Delta\alpha$. Based on the comparison between $\langle \cos^2 \theta \rangle$ for Rg_2 and that for I_2, whose polarizability anisotropy ($\Delta\alpha = 6.7 \text{Å}^3$) is well known,[57] with their different rotational constants B's corrected, $\Delta\alpha$ of Ar_2, Kr_2, and Xe_2 are successfully estimated to be 0.5, 0.7, and 1.3 Å^3, respectively. Recently, the similar technique was extended to the nonadiabatic regime. Pinkham et al. evaluated the polarizability anisotropy ($\Delta\alpha = 3.2 \text{Å}^3$) of an HBr molecule by comparing the transient alignment of HBr molecules with both the transient alignment of N_2 molecules and the theoretical predictions based on a rigid-rotor model.[58]

4.10 Probing molecular structures and dynamics by observing photoelectron angular distributions from aligned molecules

Meckel *et al.* demonstrated that the momentum distributions of the photoelectrons ejected from aligned O_2 and N_2 molecules by tunneling ionization carry the information about their HOMOs, whereas the diffraction patterns of the elastically scattered electrons reveal the interatomic distance of the molecule.[59] The molecules are aligned by a slightly stretched (60 fs) and moderately intense ($\leq 8 \times 10^{13}$ W/cm^2) pump pulse. At the well-defined time delays for alignment and anti-alignment, the molecules are ionized by a short (40 fs) and intense (2.5×10^{14} W/cm^2) probe pulse. Both the aligning (pump) and the ionizing (probe) pulses propagate collinearly along x-axis, and they are perpendicularly polarized along y- and z-axis, respectively. The photoelectrons are detected by a COLTRIMS system,[45] which allow them to measure the 3D momentum vectors of electrons and ions in coincidence. When the photoelectron momentum distributions are projected onto the three planes for display purposes, the spectra for N_2 are qualitatively indistinguishable from those of O_2. However, the structures specific to the target molecule are revealed by subtracting the anti-aligned projections from the aligned projections and dividing the difference by their sum. The resultant distributions are termed normalized differences. They find that the low lateral-momentum electrons ($p_\perp = \sqrt{p_x^2 + p_y^2} < 0.5$ au, where p is an electron momentum) show characteristic pattern in the normalized difference perpendicular to the ionizing laser field. Specifically, the very different distributions are found in the central parts of the p_x-p_y projections of the normalized differences for O_2 and N_2, which reflect the very different structures of the respective ionizing HOMOs. Here, they utilize the fact that the laser-induced tunneling serves as a filter for the perpendicular component of HOMO. They confirm that their observations are in good agreement with the theoretical calculations. On the other hand, they extract information about the scattering potential, which is related to the bond length of the diatomic molecule, by analyzing the structure at higher electron momenta ($p_\perp > 0.5$ au and $|p_z| > 1.5$ au), which are dominated by the elastically rescattered electrons. This procedure corresponds to the observation of laser-induced electron diffraction (LIED).[60] In order to remove the influence of the intense laser field on the electron momentum after the elastic scattering, they apply the analyzing procedure in analogy to the attosecond streak camera. With some appropriate assumptions,[59] they evaluate the bond length of 2.6 au for O_2, which is reasonably close to the equilibrium bond length of O_2 ($d = 2.3$ au).

Bisgaard *et al.* demonstrated ultrafast time-resolved imaging measurements of the nonadiabatic photodissociation reaction of a CS_2 molecule $CS_2 + h\nu \rightarrow CS_2^* \rightarrow CS(X) + S(^1D_2)/S(^3P_J)$ by observing photoelectron angular distributions (PADs) from aligned CS_2 molecules.[61] They first irradiate CS_2 molecules with a 100-fs and

805-nm laser pulse to induce transient alignment. During the time window of <1.5 ps at around the half-revival (73.5 ps after the alignment pulse) of molecular alignment, when CS_2 molecules are best aligned, they perform pump-probe measurements. Fixed-in-space CS_2 molecules are pumped by a 201.2-nm femtosecond pump pulse to prepare coherently a superposition of two scattering resonances, which differ only in the degree of symmetric stretch versus bending excitation. The excited CS_2 molecules are then singly ionized by a 268-nm femtosecond probe pulse at several pump-probe delays. The emitted photoelectrons are analyzed as a function of energy, 3D recoil angle, and pump-probe delay. They find that the PAD at the pump-probe delay $t = 900$ fs almost reverts back to the PAD at $t = 100$ fs, whereas the PAD at $t = 500$ fs looks different from the PADs at $t = 100$ fs and 900 fs. The reversion of the PAD results from the quantum beat interference between the two scattering resonances. They believe that the PADs at $t = 100$ fs and 900 fs reflect the average electronic character of the two scattering resonances, whereas the PAD at $t = 500$ fs reflects the difference in electronic characters between the two. In a simple model, the PADs are determined by the shape of the molecular orbital being ionized and the ionic potential from which the electron is emitted.

4.11 Observation of photoelectron angular distributions from oriented molecules with circularly polarized femtosecond pulses

Holmegaard et al. observed molecular-frame photoelectron angular distributions (MFPADs) from oriented carbonyl sulfide (OCS) and benzonitrile (C_7H_5N) molecules with intense, circularly polarized 30-fs pulses.[62] They first select the molecules in the lowest-lying rotational quantum states by an electrostatic deflector, which is crucial to achieve higher degrees of molecular orientation.[63] Then, a sample of adiabatically oriented molecules is prepared with combined weak electrostatic and intense nonresonant 10-ns laser fields. To observe the photoelectron angular distributions (PADs), a circularly polarized 30-fs probe pulse centered at ~800 nm is used.

In the case of OCS, a probe pulse at a peak intensity of 2.4×10^{14} W/cm^2 is used to ionize singly the molecules without significant fragmentation. When only the probe pulse is applied, the photoelectrons appear along the polarization plane of the probe pulse for both left and right circularly polarized (LCP and RCP) pulses. When the OCS molecules are one-dimensionally oriented with about 80% of the molecules oriented with their O atoms directed to the detector, a distinct up–down asymmetry is observed. The asymmetry reverses when the helicity of the probe pulse is flipped. For LCP (RCP) probe pulses, they find the number of photoelectrons detected in the upper part compared to the total number in the image is ~64% (39%).

In order to explain the above observations, they model the ionization process by modifying the tunneling ionization theory.[40,41] The important modification is the inclusion of Stark shifts of both OCS and OCS$^+$ energy levels, which stem from the permanent dipole and anisotropic polarizability interactions with the probe laser field. The inclusion of Stark shifts leads to an effective ionization potential, on which the tunneling ionization rate depends exponentially. The effective ionization potential shows that the oriented OCS molecule has an asymmetric ionization probability depending on whether the circularly polarized probe field has an instantaneous polarization parallel or antiparallel to the permanent dipole moment. The photoelectrons produced are then subject to the force from the remaining part of the strong probe field, which determines the final momentum distributions on the detector. Since the vector potential for the LCP (RCP) probe pulse advances the electric field by a phase of $\pi/2$ ($-\pi/2$), the forward–backward asymmetry in the tunneling ionization step is transferred to the up–down asymmetry in the final momentum distribution on the detector. They show that the theoretical calculations are in good agreement with the experimental observations.

They also observe MFPADs for asymmetric top C_7H_5N molecules. When C_7H_5N molecules are one-dimensionally oriented with a linearly polarized alignment pulse polarized parallel to the electrostatic field, the up–down asymmetry in MFPADs is observed as in the case of OCS molecules. For the LCP (RCP) probe pulse, the up/total ratio is $\sim 55\%$ ($\sim 44\%$). When C_7H_5N molecules are three-dimensionally oriented with an elliptically polarized alignment pulse with the major axis along the electrostatic field and the minor axis vertical in the image, in addition to the same up–down asymmetry as observed for one-dimensionally oriented molecules, they find that electron emission in the polarization plane, which coincides with the nodal plane of the HOMO and the HOMO-1, is suppressed.

In order to explain the above observation, they further extend the tunneling ionization theory by modeling the initial state of C_7H_5N by a simple p_x orbital with the angular node in the polarization plane and the lobes parallel to the horizontal axis. Since the presence of the nodal plane prevents electron emission in the polarization plane, electron emission occurs off-the-nodal-plane. The experimental observation is successfully reproduced by the theoretical electron momentum distribution. They also show that the theoretical off-the-nodal-plane angle $\Omega_{\text{theo}} \sim 18.8°$ is in good agreement with the experimental value $\Omega_{\text{exp}} = 18° \pm 1°$.

5. Concluding Remarks

We have reviewed the techniques for controlling the alignment and orientation of gaseous molecules based on laser technologies. The techniques can be classified into the adiabatic control, which enables the alignment and orientation in the intense laser field, and the nonadiabatic control, which enables those in the laser-field-free condition. It is often said that the nonadiabatic laser-field-free control is more advantageous than the adiabatic control because the intense laser field can modify physics and/or chemistry involved. However, we would like to emphasize that such a statement is not fair and that both the adiabatic and nonadiabatic control techniques should be employed complementarily according to the applications. In fact, both techniques have advantages and disadvantages. The adiabatic control techniques can keep the molecular alignment and orientation for a relatively long time of typically a few nanoseconds and can achieve high degrees of alignment and orientation especially for relatively heavy molecules with sufficiently low rotational temperatures though, as mentioned above, the intense laser field can modify physics and/or chemistry involved. On the other hand, although the nonadiabatic control techniques can achieve the molecular alignment and orientation in the laser-field-free condition, the degrees of alignment and orientation are generally modest and the time during which the molecular alignment and orientation can be maintained is typically shorter than one picosecond.

In general, higher degrees of alignment and orientation are more advantageous in applications. When thermal ensembles are used, it is of crucial importance to make the initial rotational temperature as low as possible. A recently developed Even–Lavie type pulsed valve can be a powerful tool to that end.[64] An alternative and complementary approach is to use rotationally state-selected molecules, which correspond to the rotational temperature of 0 K. A hexapole focusing apparatus[10] and a molecular deflector[63] can be used to select specific rotational states of molecules. With a sample of rotationally state-selected molecules, a fairly high degree of alignment and orientation could be achieved even with the combination of a modest electrostatic field and a simple femtosecond laser field[65] or with a simple femtosecond two-color laser field.[38,39] A sample of aligned molecules has already come into practical use. We hope that a sample of oriented molecules will also be used in various applications and that many interesting phenomena will be explored and utilized using an anisotropic quantum system with the broken inversion symmetry.

References

1. Special issue on Stereodynamics of Chemical Reactions, *J. Phys. Chem. A* **101**, 7461 (1997).
2. D. Herschbach, *Eur. Phys. J. D* **38**, 3 (2006).
3. D. Zeidler, A. Staudte, A. B. Bardon, D. M. Villeneuve, R. Dörner and P. B. Corkum, *Phys. Rev. Lett.* **95**, 203003 (2005).

4. J. Itatani, J. Levesque, D. Zeidler, H. Niikura, H. Pépin, J. C. Kieffer, P. B. Corkum and D. M. Villeneuve, *Nature* (London) **432**, 867 (2004).
5. T. Kanai, S. Minemoto and H. Sakai, *Nature* (London) **435**, 470 (2005).
6. T. Kanai, S. Minemoto and H. Sakai, *Phys. Rev. Lett.* **98**, 053002 (2007).
7. P. B. Corkum, *Phys. Rev. Lett.* **71**, 1994 (1993).
8. B. Friedrich and D. R. Herschbach, *Nature* (London) **353**, 412 (1991).
9. H. J. Loesch and A. Remscheid, *J. Phys. Chem.* **95**, 8194 (1991).
10. V. A. Cho and R. B. Bernstein, *J. Phys. Chem.* **95**, 8129 (1991).
11. B. Friedrich and D. Herschbach, *Phys. Rev. Lett.* **74**, 4623 (1995).
12. B. Friedrich and D. Herschbach, *J. Phys. Chem.* **99**, 15686 (1995).
13. H. Sakai, C. P. Safvan, J. J. Larsen, K. M. Hilligsøe, K. Hald and H. Stapelfeldt, *J. Chem. Phys.* **110**, 10235 (1999).
14. A. T. J. B. Eppink and D. H. Parker, *Rev. Sci. Instrum.* **68**, 3477 (1997).
15. J. J. Larsen, H. Sakai, C. P. Safvan, I. Wendt-Larsen and H. Stapelfeldt, *J. Chem. Phys.* **111**, 7774 (1999).
16. J. J. Larsen, I. Wendt-Larsen and H. Stapelfeldt, *Phys. Rev. Lett.* **83**, 1123 (1999).
17. J. J. Larsen, K. Hald, N. Bjerre, H. Stapelfeldt and T. Seideman, *Phys. Rev. Lett.* **85**, 2470 (2000).
18. J. Karczmarek, J. Wright, P. Corkum and M. Ivanov, *Phys. Rev. Lett.* **82**, 3420 (1999).
19. D. M. Villeneuve, S. A. Aseyev, P. Dietrich, M. Spanner, M. Yu. Ivanov and P. B. Corkum, *Phys. Rev. Lett.* **85**, 542 (2000).
20. H. Stapelfeldt and T. Seideman, *Rev. Mod. Phys.* **75**, 543 (2003).
21. T. Seideman, *J. Chem. Phys.* **103**, 7887 (1995).
22. F. Rosca-Pruna and M. J. J. Vrakking, *Phys. Rev. Lett.* **87**, 153902 (2001).
23. T. Seideman and E. Hamilton, *Adv. At. Mol. Opt. Phys.* **52**, 289 (2005).
24. T. Suzuki, Y. Sugawara, S. Minemoto and H. Sakai, *Phys. Rev. Lett.* **100**, 033603 (2008).
25. K. F. Lee, D. M. Villeneuve, P. B. Corkum, A. Stolow and J. G. Underwood, *Phys. Rev. Lett.* **97**, 173001 (2006).
26. B. Friedrich and D. Herschbach, *J. Chem. Phys.* **111**, 6157 (1999).
27. B. Friedrich and D. Herschbach, *J. Phys. Chem. A* **103**, 10280 (1999).
28. H. Sakai, S. Minemoto, H. Nanjo, H. Tanji and T. Suzuki, *Phys. Rev. Lett.* **90**, 083001 (2003).
29. S. Minemoto, H. Nanjo, H. Tanji, T. Suzuki and H. Sakai, *J. Chem. Phys.* **118**, 4052 (2003).
30. H. Tanji, S. Minemoto and H. Sakai, *Phys. Rev. A* **72**, 063401 (2005).
31. M. Machholm and N. E. Henriksen, *Phys. Rev. Lett.* **87**, 193001 (2001).
32. M. Spanner, E. A. Shapiro and M. Ivanov, *Phys. Rev. Lett.* **92**, 093001 (2004).
33. Y. Sugawara, A. Goban, S. Minemoto and H. Sakai, *Phys. Rev. A* **77**, 031401(R) (2008).
34. A. Goban, S. Minemoto and H. Sakai, *Phys. Rev. Lett.* **101**, 013001 (2008).
35. J. G. Underwood, M. Spanner, M. Y. Ivanov, J. Mottershead, B. J. Sussman and A. Stolow, *Phys. Rev. Lett.* **90**, 223001 (2003).
36. T. Kanai and H. Sakai, *J. Chem. Phys.* **115**, 5492 (2001).
37. K. Oda, M. Hita, S. Minemoto and H. Sakai, *Phys. Rev. Lett.* **104**, 213901 (2010).
38. M. Muramatsu, M. Hita, S. Minemoto and H. Sakai, *Phys. Rev. A* **79**, 011403(R) (2009).

39. S. De, I. Znakovskaya, D. Ray, F. Anis, N. G. Johnson, I. A. Bocharova, M. Magrakvelidze, B. D. Esry, C. L. Cocke, I. V. Litvinyuk and M. F. Kling, *Phys. Rev. Lett.* **103**, 153002 (2009).
40. X. M. Tong, Z. X. Zhao and C. D. Lin, *Phys. Rev. A* **66**, 033402 (2002).
41. M. V. Ammosov, N. B. Delone and V. P. Krainov, *Zh. Eksp. Teor. Fiz.* **91**, 2008 (1986) [*Sov. Phys. JETP* **64**, 1191 (1986)].
42. I. V. Litvinyuk, K. F. Lee, P. W. Dooley, D. M. Rayner, D. M. Villeneuve and P. B. Corkum, *Phys. Rev. Lett.* **90**, 233003 (2003).
43. P. W. Dooley, I. V. Litvinyuk, K. F. Lee, D. M. Rayner, M. Spanner, D. M. Villeneuve and P. B. Corkum, *Phys. Rev. A* **68**, 023406 (2003).
44. X.-M. Tong and C. D. Lin, unpublished.
45. J. Ullrich, R. Moshammer, A. Dorn, R. Dörner, L. Ph. H. Schmidt and H. Schmidt-Böcking, *Rep. Prog. Phys.* **66**, 1463 (2003).
46. T. Suzuki, S. Minemoto, T. Kanai and H. Sakai, *Phys. Rev. Lett.* **92**, 133005 (2004).
47. Y. Silberberg, *Nature* (London) **430**, 624 (2004).
48. T. Kanai, S. Minemoto and H. Sakai, *Ultrafast Phenomena XV*, P. Corkum, D. Jonas, R. J. Miller and A. M. Weiner (Eds.), (Springer, 2007), pp. 27–29.
49. A. C. Kak and M. Slaney, *Principles of Computerized Tomographic Imaging* (Society for Industrial and Applied Mathematics, New York, 2001).
50. J. P. Marangos, *Nature* (London) **435**, 435 (2005).
51. M. Lein, *J. Phys. B* **40**, R135 (2007).
52. S. Minemoto, T. Kanai and H. Sakai, *Phys. Rev. A* **77**, 041401(R) (2008).
53. Y. Sato, H. Kono, S. Koseki and Y. Fujimura, *J. Am. Chem. Soc.* **125**, 8019 (2003).
54. Y. Fujimura, L. González, K. Hoki, J. Manz and Y. Ohtsuki, *Chem. Phys. Lett.* **306**, 1 (1999).
55. P. Lan, P. Lu, W. Cao, Y. Li and X. Wang, *Phys. Rev. A* **76**, 021801(R) (2007).
56. S. Minemoto, H. Tanji and H. Sakai, *J. Chem. Phys.* **119**, 7737 (2003).
57. D. W. Callahan, A. Yokozeki and J. S. Muenter, *J. Chem. Phys.* **72**, 4791 (1980).
58. D. Pinkham, T. Vogt and R. R. Jones, *J. Chem. Phys.* **129**, 064307 (2008).
59. M. Meckel, D. Comtois, D. Zeidler, A. Staudte, D. Pavičić, H. C. Bandulet, H. Pépin, J. C. Kieffer, R. Dörner, D. M. Villeneuve and P. B. Corkum, *Science* **320**, 1478 (2008).
60. T. Zuo, A. Bandrauk and P. B. Corkum, *Chem. Phys. Lett.* **259**, 313 (1996).
61. C. Z. Bisgaard, O. J. Clarkin, G. Wu, A. M. D. Lee, O. Geßner, C. C. Hayden and A. Stolow, *Science* **323**, 1464 (2009).
62. L. Holmegaard, J. L. Hansen, L. Kalhøj, S. L. Kragh, H. Stapelfeldt, F. Filsinger, J. Küpper, G. Meijer, D. Dimitrovski, M. Abu-samha, C. P. J. Martiny and L. B. Madsen, *Nature Phys.* **6**, 428 (2010).
63. L. Holmegaard, J. H. Nielsen, I. Nevo, H. Stapelfeldt, F. Filsinger, J. Küpper and G. Meijer, *Phys. Rev. Lett.* **102**, 023001 (2009).
64. U. Even, J. Jortner, D. Noy, N. Lavie and C. Cossart-Magos, *J. Chem. Phys.* **112**, 8068 (2000).
65. O. Ghafur, A. Rouzée, A. Gijsbertsen, W. K. Siu, S. Stolte and M. J. J. Vrakking, *Nature Phys.* **5**, 289 (2009).

Chapter 5

Electronic and Nuclear Dynamics in Intense Laser Fields

In this chapter, electronic and nuclear dynamics of molecules in intense laser fields are presented. First, electron–nuclei correlated motions of a hydrogen molecular ion are elucidated based on the quantum mechanical treatment. Concept of phase-adiabatic states is introduced to describe electronic and nuclear dynamics of molecules in intense laser fields. Second, a theoretical treatment of interelectronic correlations in a hydrogen molecule under an intense laser field conditions is presented. Third, experimental and theoretical studies of ionization and fragmentation of carbon dioxide, benzene, and fullerene in intense infrared laser pulses are reviewed. Finally, experimental results of proton dynamics in polyatomic molecules are shown.

1. Electron–Nuclei Correlated Motions in H_2^+

In Secs. 4 to 9 of Chap. 1, electronic and nuclear dynamics of molecules in intense laser fields were surveyed. In this section, a quantum dynamical treatment of a hydrogen molecular ion in intense laser fields is presented for quantitative explanations of correlated dynamics between the electron and nuclei. First, the electronic and nuclear Hamiltonian is expressed in terms of cylindrical coordinates. The coordinates are appropriate for homonuclear diatomic molecules in a linearly polarized laser. Next, phase-adiabatic states are introduced as the basis set of molecules in an intense laser field. Its adiabatic parameter is the phase of the laser. Electron transfer between the two-electronic potential well of H_2^+ is demonstrated as well as the ionization by using a realistic three-dimensional (3D) model. A slowdown of nuclear motions in the dissociative ionization process is presented as one of the correlated electronic and nuclear motions.

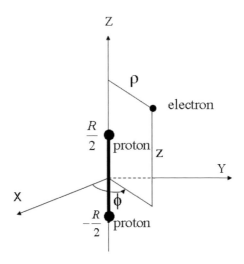

Fig. 1. Hydrogen molecular ion in cylindrical coordinate representation (ρ, z, ϕ). A linearly polarized electric field of a laser pulse is applied along the Z-axis. The nuclear motion is restricted to the Z-axis. R denotes the internuclear distance. ρ and z have their origin at the center of mass of the two nuclei r_{cn}. The electron motion is described in terms of two coordinates, ρ and z, since H_2^+ has cylindrical symmetry.

1.1 Hamiltonian in terms of cylindrical coordinates

To describe the dynamics of H_2^+ in a linearly polarized laser, we use cylindrical coordinates (ρ, z, ϕ) as shown in Fig. 1.[1] The time-dependent Schrödinger equation for the relative motion of H_2^+ can be expressed in atomic units after separation of the center of mass coordinate as

$$i\frac{\partial}{\partial t}\psi(\rho, z, R) = [T_n(R) + T_e(\rho, z) + U(\rho, z, R) + V_E(z, t)]\psi(\rho, z, R). \quad (1)$$

Here, $T_n(R)$ is the nuclear kinetic energy operator and $T_e(\rho, z)$ is the electronic kinetic energy operator.

The nuclear kinetic energy operator is expressed as

$$T_n(R) = -\frac{1}{m_p}\frac{\partial^2}{\partial R^2}, \quad (2)$$

where m_p is the proton mass.

The electron kinetic energy operator is expressed as

$$T_e(\rho, z) = -\frac{1}{2\mu}\left(\frac{\partial^2}{\partial \rho^2} + \frac{1}{\rho}\frac{\partial}{\partial \rho} + \frac{\partial^2}{\partial z^2}\right), \quad (3)$$

where $\mu = 2m_p m_e/(2m_p + m_e)$. Electron rotation with the polar angle ϕ is independent of the electric field of a linearly polarized laser along the z-axis, and rotational energy is excluded in the kinetic energy operator. m is the quantum number of the electronic angular momentum operator $-i d/d\phi$. Therefore, the Coulomb potential $U(\rho, z, R)$ is given as a function of ρ, z, and R as

$$U(\rho, z, R) = \frac{m^2}{2\rho^2} + \frac{1}{R} - \frac{1}{\sqrt{\rho^2 + \left(z - \frac{R}{2}\right)^2}} - \frac{1}{\sqrt{\rho^2 + \left(z + \frac{R}{2}\right)^2}}. \quad (4)$$

The dipole interaction between H_2^+ and a laser pulse of electric field $E(t)$, $V_E(z, t)$, is given as

$$V_E(z, t) = z\left(1 + \frac{m_e}{2m_p + m_e}\right) E(t). \quad (5)$$

Here, the electric dipole moment is defined as the sum of the electric dipole moment from the center of mass (z_c) to each charged particle of H_2^+. That is, $(z_a - z_c) + (z_b - z_c) - (z_e - z_c) = -2(z_c - z_{cn}) - [z_e - z_{cn} - (z_c - z_{cn})]$, where $z_c = [m_e z_e + m_p(z_a + z_b)]/(m_e + 2m_p)$, the center of mass of the two protons is $z_{cn} = (z_a + z_b)/2$ and $z_c - z_{cn} = z m_e/(M_e + 2m_p)$ with $z = z_c - z_{cn}$. Here, subscripts a (b) and e denote two nuclei and the electron, respectively.

The electric field of laser $E(t)$ is assumed to have the form $E(t) = E_0 \sin(\pi t/T) \sin(\omega t)$ for $0 \leq t \leq T$ and $E(t) = 0$ for $t < 0$ or $T < t$, where E_0 is amplitude, T is pulse duration, and ω is carrier frequency. The laser is set to be parallel to the molecular axis.

We are interested in the dynamic behaviors of a hydrogen molecular ion in an intense laser field in which electronic potential is strongly distorted. Figure 2 shows the effective Coulomb potential $U(\rho, z, R) + V_E(z, t)$ with t fixing at 224 (5.6 fs) as a function of z with the other coordinates being fixed, $U(\rho = 0.01, z, R = \text{const}) + V_E(z, t)$ of H_2^+. The bare Coulomb potential energy at $R = 2.0$ (1.1 Å) is shown in Fig. 2(a). $R = 3.0$ (1.6 Å), $R = 4.0$ (2.1 Å), and $R = 8.0$ (4.2 Å) at $t = 224$ (5.6 fs) are shown in Figs. 2(b), 2(c), and 2(d), respectively. The laser parameters are set to $E_0 = 0.096$ (3.2×10^{14} W/cm^2), $T = 400$ (10 fs), and $\omega = 0.0515$ (884 nm). In Fig. 2, two sinks correspond to the position at which the two protons are located. It can be seen that the effective potential is strongly distorted by the Coulomb potential: two barriers, inner and outer, are created. The height of the inner barrier peak depends on the distance between the core ions. At $R = 4.0$, the height is almost equal to that of the outside wall, at which so-called enhanced ionization is induced.

The time-dependent Schrödinger equation (1) was solved by using the finite-difference scheme described in Chap. 3. In actual calculations, a generalized cylindrical coordinate system, which is introduced in Chap. 3, was employed for the

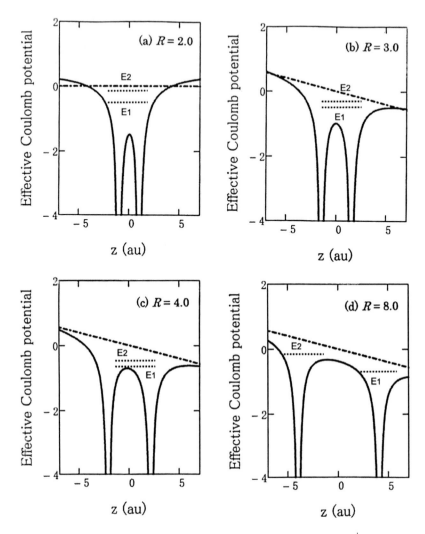

Fig. 2. Snapshots of effective Coulomb potential $U(\rho, z, R) + V_E(z, t)$ of H_2^+ as a function of z at $\rho = 0.01$: (a) $R = 2.0$ (1.1 Å) at $t = 0$, (b) $R = 3.0$ (1.6 Å) at $t = 224$ (5.6 fs), (c) $R = 4.0$ (2.1 Å) at $t = 224$ (5.6 fs), and (d) $R = 8.0$ (4.2 Å) at $t = 224$ (5.6 fs). The dotted-broken lines denote the molecule–laser interaction energy, $V_E(z, t)$. Dotted lines denote potential energies E_1 and E_2 of phase-adiabatic states, 1 and 2, at each R and t, respectively.

finite-difference scheme to give sufficient accuracy to the cylindrical coordinates. As the basis set for the Born–Oppenheimer electronic states, $1\sigma_g$ and $1\sigma_u$ states were adopted. These two states are the charge resonance states that are strongly coupled to each other by the electric field of the laser.[2] Therefore, these two states constitute a fairly good basis set for describing the optically induced electron dynamics of H_2^+, although the two-electronic state model cannot directly treat the ionization process.

1.2 Phase-adiabatic states

The phase-adiabatic electronic eigenfunctions, $|1\rangle$ and $|2\rangle$, are expressed in terms of the $1s\sigma_g$ (abbreviated as g) and $2p\sigma_u$ (abbreviated as u) states as

$$|1\rangle = \cos\theta(t)|g\rangle - \sin\theta(t)|u\rangle, \tag{6a}$$

and

$$|2\rangle = \sin\theta(t)|g\rangle + \cos\theta(t)|u\rangle, \tag{6b}$$

where

$$\tan 2\theta(t) = \frac{2\langle g|z|u\rangle E(t)}{\Delta E_{ug}(R)}. \tag{7}$$

Here, $\Delta E_{ug}(R)$ is the energy separation between $1s\sigma_g$ and $2p\sigma_u$ states,

$$\Delta E_{ug}(R) = E_u(R) - E_g(R). \tag{8}$$

The adiabatic energies are expressed as

$$E_{1,2}(R,t) = \frac{1}{2}\left\{E_g(R) + E_u(R) \mp \sqrt{\Delta E_{ug}(R)^2 + 4|\langle g|z|u\rangle E(t)|^2}\right\}. \tag{9}$$

These are the solution of the electronic Hamiltonian in the presence of a laser field.

Two adiabatic parameters, R and t, are included in Eq. (9): one is the nuclear coordinate and the other is the phase of the applied laser field ($E(t)$). The adiabatic states, Eqs. (6a, b), are called *phase-adiabatic states* to emphasis the adiabaticity with respect to the oscillating electric field of the laser.

The phase-adiabatic potential energies of H_2^+ as a function of R are illustrated in Fig. 3. Here, the same parameters as those in Fig. 2 were adopted for the pulsed laser. The phase-adiabatic potential energies $E_1(R,t)$ and $E_2(R,t)$ are equal to the Born–Oppenheimer adiabatic potential energies $E_g(R)$ and $E_u(R)$, respectively, at $t = n\pi/\omega$ ($n = 1, 2, 3, \ldots$) when the electric field $E(t) = E_0 \sin(\omega t) = 0$.

The total wavefunction is expanded in terms of the phase-adiabatic states within the two-state model as

$$|\psi\rangle = \chi_1(R)|1\rangle + \chi_2(R)|2\rangle, \tag{10}$$

where expansion coefficients, $\chi_1(R)$ and $\chi_2(R)$, are the nuclear wavefunctions of the adiabatic states, $|1\rangle$ and $|2\rangle$, respectively.

After substitution of Eq. (10) into Eq. (1), the nuclear wavefunctions, $\chi_1(R)$ and $\chi_2(R)$, satisfy the coupled equations:

$$i\frac{\partial}{\partial t}\chi_1(R) = \left[-\frac{1}{m_p}\frac{\partial^2}{\partial R^2} + E_1(R,t) + \frac{1}{m_p}\left(\frac{\partial\theta}{\partial R}\right)^2\right]\chi_1(R)$$

$$- \Lambda(R,t)\chi_2(R) \tag{11a}$$

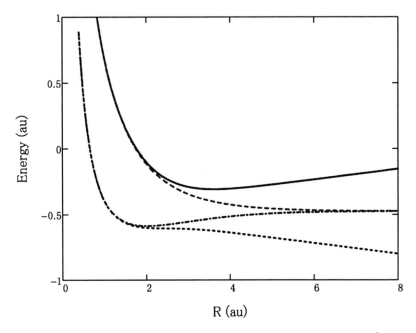

Fig. 3. Phase-adiabatic potential energies, E_1 (dotted line) and E_2 (solid line), of H_2^+ as a function of the internuclear distance R. The dotted-broken and broken lines denote the potential energies (E_g and E_u) of $1s\sigma_g$ and $2p\sigma_u$, respectively. The same values as those in Fig. 2 were adopted for variables of the pulsed laser.

and

$$i\frac{\partial}{\partial t}\chi_2(R) = \left[-\frac{1}{m_p}\frac{\partial^2}{\partial R^2} + E_2(R,t) + \frac{1}{m_p}\left(\frac{\partial \theta}{\partial R}\right)^2\right]\chi_2(R)$$
$$+ \Lambda(R,t)\chi_1(R), \qquad (11b)$$

where $\Lambda(R,t)$, nonadiabatic coupling operator, is given as

$$\Lambda(R,t) = i\frac{\partial \theta}{\partial t} + \frac{2}{m_p}\frac{\partial \theta}{\partial R}\frac{\partial}{\partial R} + \frac{1}{m_p}\frac{\partial^2 \theta}{\partial R^2} + \frac{2}{m_p}\left\langle g\left|\frac{\partial}{\partial R}\right|u\right\rangle\frac{\partial}{\partial R}. \qquad (12)$$

In deriving Eq. (12), the diagonal terms of nonadiabatic kinetic energy corrections, $\langle g|\frac{\partial^2}{\partial R^2}|g\rangle/m_p$ and $\langle u|\frac{\partial^2}{\partial R^2}|u\rangle/m_p$, which lead to energy shifts in the adiabatic potentials, were omitted as in the ordinary treatment of nonadiabatic couplings. The nonadiabatic term $\langle g|\frac{\partial^2}{\partial R^2}|u\rangle/m_p$ was also neglected.

The nonadiabatic coupling operator in Eq. (12) consists of two types of nonadiabatic couplings: one is the first term that originates from phase change in the electric field of the laser pulse and the other is all of the other terms that originate from the nuclear kinetic energy operator. The former is the laser-induced nonadiabatic coupling and the latter is breakdown of the Born–Oppenheimer approximation.

Field-induced nonadiabatic coupling has a dominant contribution and breakdown of the Born–Oppenheimer approximation has a minor contribution under intense laser field conditions. Therefore, by omitting the nonadiabatic coupling operators due to the nuclear motions, the coupling operator is approximately written as

$$\Lambda(R, t) \approx i\frac{\partial \theta(t)}{\partial t}. \tag{13}$$

It can easily be shown that $\langle 1|i\hbar\frac{\partial}{\partial t}|2\rangle = i\hbar\frac{\partial \theta(t)}{\partial t}$, i.e., the operator of the nonadiabatic transition between two phase-adiabatic states is time-derivative, $i\hbar\frac{\partial}{\partial t}$.

The laser-induced nonadiabatic coupling is expressed from Eq. (7) as

$$i\hbar\frac{\partial \theta(t)}{\partial t} = \frac{i\hbar \Delta E_{\text{ug}}\langle g|z|u\rangle\frac{d}{dt}E(t)}{\Delta E_{\text{ug}}^2 + 4|\langle g|z|u\rangle|^2 E(t)^2}. \tag{14}$$

If the magnitude of the laser-induced nonadiabatic coupling is less than the difference in the phase-adiabatic energies between the two corresponding electronic states, the adiabatic approximation is a good approximation. On the other hand, if the magnitude becomes equal to or larger than the energy difference, the adiabatic approximation breaks down. From the above consideration, it can be seen that a measure of the adiabaticity is defined by

$$\delta = \left|\frac{\hbar\frac{\partial \theta(t)}{\partial t}}{E_2(R, t) - E_1(R, t)}\right|. \tag{15}$$

This is called (phase) adibaticity parameter in this book.

Adiabatic approximation is valid for $\delta \ll 1$, while adiabatic approximation breaks down for $\delta \geq 1$.

When $\Delta E_{\text{ug}} \gg \langle g|z|u\rangle E(t)$ is satisfied, Eq. (15) is approximated as

$$\delta \approx \left|\frac{\hbar\langle g|z|u\rangle\frac{d}{dt}E(t)}{\Delta E_{\text{ug}}^2}\right|. \tag{16}$$

In addition, when the electric field of the pulse laser has a form $E(t) = E_0 \sin(\pi t/T) \sin(\omega t)$ with $\omega T \gg 1$, the time derivative of the pulse envelope can be neglected, and Eq. (16) can be written as

$$\delta \approx \left|\frac{\hbar\omega\langle g|z|u\rangle E_0 \cos(\omega t)}{\Delta E_{\text{ug}}^2}\right|. \tag{17}$$

Here, the transition dipole matrix element is given as $\langle g|z|u\rangle \simeq \frac{R}{2}.$[3]

1.3 Interwell electron transfer

An interesting electronic dynamics takes place within one cycle (3 fs) of the laser as shown in Fig. 4. This is the electron transfer between two nuclei of H_2^+. Let

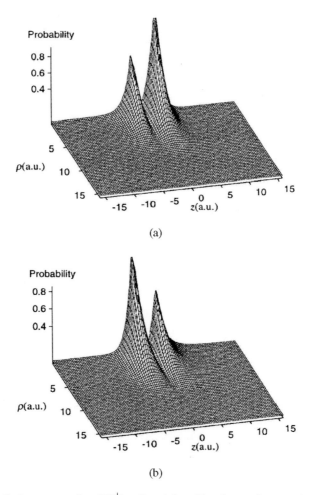

Fig. 4. Interwell electron transfer of H_2^+ at $R = 4.0$ au. The electron is set to the upper electronic state $2p\sigma_u$ at $t = 0$. A laser pulse of electric field $E(t) = E_0 \sin(\pi t/T) \sin(\omega t)$ with $E_0 = 0.096$ (3.2×10^{14} W/cm^2), $T = 400$ (10 fs) and $\omega = 884$ nm is applied to the electron. (a) 2D plot of the electron probability at $t = 31$ (0.75 fs) with $E(t) = 0.023$ (1.9×10^{13} W/cm^2) and central frequency of the laser pulse and (b) 2D plot of the electron probability at $t = 92$ (2.2 fs) with $E(t) = -0.063$ (1.4×10^{14} W/cm^2). These plots show the difference in the electron behaviors between a half period of the laser oscillation. (Reprinted with permission from Kawata et al.[1(b)] Copyright (1999) by American Institute of Physics).

us call this interwell electron transfer. In Fig. 4, the electron densities are plotted as a function of ρ and z at $R = 4.0$. Here, the electron was set be in the excited electronic state $2p\sigma_u$ at $t = 0$. The laser parameters were assumed to be the same as those in Fig. 2. Figures 4(a) and 4(b) show the electron densities at $t = 31$ (0.75 fs) with $E(t) = 0.023$ (1.9×10^{13} W/cm^2) and at $t = 92$ (2.2 fs) with $E(t) = -0.063$ (5.2×10^{13} W/cm^2), respectively.[1] The difference between the two times, 61 (1.5 fs), corresponds to the half cycle of the applied laser with $\omega = 884$ nm. The asymmetric electron densities in each figure and the mirror image between the two figures are due to the interwell electron transfer. That is, the large electron density probability in Fig. 4(a) originates from the electron transfer from the left ion core to the right ion core. In Fig. 4(b), on the other hand, the reverse electron transfer is induced. Thus, the electron moves between the two ion cores on the upper phase-adiabatic potential. Interwell electron transfer effectively takes place when the adiabaticity condition is satisfied. Actually, the magnitude of the adiabaticity parameter δ is an order of 0.01 in Fig. 4(a) at $t = 31$ and $R = 4.0$, and adiabaticity is almost maintained. In Fig. 4(b) depicted at $t = 92$ and $R = 4.0$, δ is an order of 0.1. The adiabaticity condition is still fulfilled for Fig. 4(b).

Figure 5 shows results of a 3D calculation of time-dependent populations under the same excitation condition as that in Fig. 4. In Fig. 5, the nuclear wavepacket motions are taken into account. Here, populations of the lower and upper phase-adiabatic states, $|1\rangle$ and $|2\rangle$, and the localized state at the left ion core, $|L\rangle$, are plotted. It can be seen from Fig. 5 that the system behaves adiabatically at the initial stage, *i.e.*, in the first half cycle of the pulse ($0 < t \leq 61$). In other words, the system stays in the adiabatic state $|2\rangle$. In this region, the population of $|L\rangle$ is decreased. This decrease originates from the interwell electron transfer from $|L\rangle$ to $|R\rangle$.

In the second half cycle ($61 < t \leq 122$), the population of $|L\rangle$ is around 0.9, which means reverse interwell electron transfer from $|R\rangle$ to $|L\rangle$. This shows that the electron transfer follows adiabatically the electric field of the laser pulse. Population changes in the adiabatic states can be seen, *i.e.*, decrease in $|2\rangle$ and increase in $|1\rangle$.

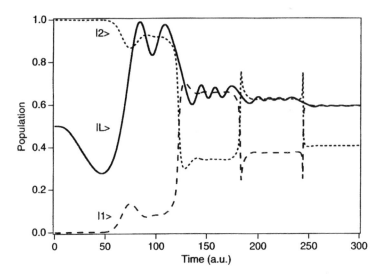

Fig. 5. Time-dependent populations of H_2^+. Solid, broken, and dotted lines denote the populations of states, $|L\rangle$, $|1\rangle$, and $|2\rangle$, respectively. (Reprinted with permission from Kawata *et al.*[1(b)] Copyright (1999) by American Institute of Physics).

This is due to breakdown of the adiabatic approximation. In addition, an oscillatory behavior appears in $|L\rangle$. This is due to interferences between the two adiabatic states.

To qualitatively understand the oscillatory behaviors shown in Fig. 5, let us introduce a localized basis set (denoted by R and L) defined as

$$|R\rangle = \frac{1}{\sqrt{2}}(|g\rangle + |u\rangle) \tag{18a}$$

and

$$|L\rangle = \frac{1}{\sqrt{2}}(|g\rangle - |u\rangle), \tag{18b}$$

where $|R\rangle$ denotes the ket vector of the electronic state localized near the right nuclei of H_2^+ and $|L\rangle$ denotes the ket vector of an electronic state localized near the left nuclei.

The populations in the left and right wells, $P_L(R)$ and $P_R(R)$, are respectively expressed as

$$\begin{aligned}P_L(R) &= |\langle L|\psi\rangle|^2 \\ &= \sin^2\left(\theta + \frac{\pi}{4}\right)|\chi_1(R)|^2 + \cos^2\left(\theta + \frac{\pi}{4}\right)|\chi_2(R)|^2 \\ &\quad - \text{Re}\{\cos(2\theta)\chi_1^*(R)\chi_2(R)\},\end{aligned} \tag{19a}$$

and

$$\begin{aligned}P_R(R) &= |\langle R|\psi\rangle|^2 \\ &= \cos^2\left(\theta + \frac{\pi}{4}\right)|\chi_1(R)|^2 + \sin^2\left(\theta + \frac{\pi}{4}\right)|\chi_2(R)|^2 \\ &\quad + \text{Re}\{\cos(2\theta)\chi_1^*(R)\chi_2(R)\}.\end{aligned} \tag{19b}$$

Here, the third term in each equation represents the interference between two phase-adiabatic states, $|1\rangle$ and $|2\rangle$. The conservation of the system is given as $P_L(R) + P_R(R) = |\chi_1(R)|^2 + |\chi_2(R)|^2$.

In the first half cycle of the laser pulse, the system remains in adiabatic state $|2\rangle$. In this condition, populations of $|L\rangle$ and $|R\rangle$ states are expressed as

$$P_L(R) \approx \cos^2\left(\theta + \frac{\pi}{4}\right)|\chi_2(R)|^2 \quad \text{and} \quad P_R(R) \approx \sin^2\left(\theta + \frac{\pi}{4}\right)|\chi_2(R)|^2$$

with $\theta \approx 0$. The magnitude of θ increases after the pulse is turned on, and $P_L(R) < P_R(R)$. In the second half cycle, the interference terms in Eq. (19) begin to make a contribution to population changes. The interference term can be rewritten in terms of its amplitude and phase as

$$\cos(2\theta)\chi_1^*(R)\chi_2(R)$$
$$\sim \cos(2\theta)|\chi_1^*(R)\chi_2(R)|\exp\left[-i\int dt'(E_2(t')-E_1(t'))\right].$$

The phase term explains the oscillation that appears in Fig. 5: the period of the phase term, $2\pi/[E_2(t)-E_1(t)] \sim 23$ at $t=96$, is close to the difference between the two peaks, $t=84$ and 108.

The electron transfer probability decreases, as the internuclear bond stretches. This is because the adiabaticity condition breaks down and the nonadiabatic transition probability between the two phase-adiabatic states increases.

So far, we have discussed the electron dynamics from the quantum mechanical viewpoint. It is also interesting to consider the electron dynamics from the semiclassical viewpoint since the electron transfer probability is easily estimated. In the semiclassical treatment, electron motions are treated quantum mechanically, while nuclear motions are treated classically. There are several references for semiclassical treatments including the Landau-Zener method.[4]

Let us consider a 1D motion in a two-state model. The potentials are schematically shown in Fig. 6.

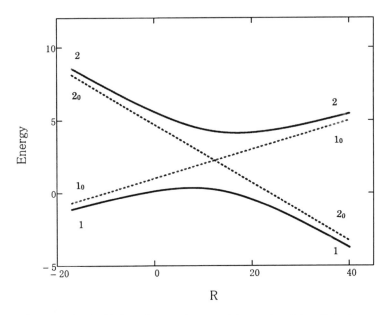

Fig. 6. 1D linear model. Solid lines denotes the potential energies of the adiabatic states, 1 and 2, and dotted lines denote the potential energies of the zero-order states, 1_0 and 2_0.

In this model, the two adiabatic states $|1\rangle$ and $|2\rangle$ are expanded in terms of zero-order states, $|1_0\rangle$ and $|2_0\rangle$, called diabatic states as

$$|1\rangle = \cos\theta|1_0\rangle - \sin\theta|2_0\rangle, \qquad (20a)$$

$$|2\rangle = \sin\theta|1_0\rangle + \cos\theta|2_0\rangle. \qquad (20b)$$

The Hamiltonian matrix is expressed within a linear approximation as

$$\begin{pmatrix} E_0 + \dfrac{\Delta F}{2}\Delta R & \gamma \\ \gamma & E_0 - \dfrac{\Delta F}{2}\Delta R \end{pmatrix}. \qquad (21)$$

Here, $\Delta F = -\frac{\partial}{\partial R}(H_{1_0 1_0} - H_{2_0 2_0})|_{R=R_c}$, $\Delta R = R - R_c$, and $\gamma = H_{1_0 2_0}(R_c)$. The adiabatic potential energies are expressed as $E_{1,2} = E_0 \mp \frac{1}{2}\sqrt{(\Delta F \Delta R)^2 + 4\gamma^2}$ with $E_0 = H_{1_0 1_0}(R_c) = H_{2_0 2_0}(R_c)$. The coefficients of Eq. (20) have the relation $\tan 2\theta = \frac{2\gamma}{\Delta F \Delta R}$.

The Landau–Zener formula for the nonadiabatic transition probability from $|1\rangle$ to $|2\rangle$, P_{12}, is given as[5]

$$P_{12} = \exp\left[-\frac{2\pi\gamma^2}{\hbar|\Delta F v|}\right], \qquad (22)$$

where $v = \frac{d(\Delta R)}{dt}$.

The H_2^+ energy matrix is expressed in terms of the $|g\rangle$ and $|u\rangle$ basis set as

$$\begin{pmatrix} \dfrac{\Delta E_{gu}}{2} & \langle g|z|u\rangle E(t) \\ \langle u|z|g\rangle E(t) & -\dfrac{\Delta E_{gu}}{2} \end{pmatrix}. \qquad (23)$$

Here, R is fixed and the origin of the energy is set to $(E_g + E_u)/2$.

Phase-adiabatic states $|1\rangle$ and $|2\rangle$ are expressed from Eqs. (6) and (18) in terms of the localized basis set as

$$|1\rangle = \cos\left(\theta(t) - \frac{\pi}{4}\right)|L\rangle - \sin\left(\theta(t) - \frac{\pi}{4}\right)|R\rangle \qquad (24a)$$

and

$$|2\rangle = \sin\left(\theta(t) - \frac{\pi}{4}\right)|L\rangle + \cos\left(\theta(t) - \frac{\pi}{4}\right)|R\rangle. \qquad (24b)$$

Here, the coefficients satisfy the equation

$$\tan\left(2\theta(t) - \frac{\pi}{2}\right) = -\cot 2\theta(t) = -\frac{\Delta E_{ug}(R)}{2\langle g|z|u\rangle E(t)}. \qquad (25)$$

It can be seen from Eq. (25) that the energy matrix is expressed in terms of the localized basis set as

$$\begin{pmatrix} \langle g|z|u\rangle E(t) & \dfrac{\Delta E_{gu}}{2} \\ \dfrac{\Delta E_{gu}}{2} & -\langle g|z|u\rangle E(t) \end{pmatrix}. \qquad (26)$$

This expression shows that transitions in the localized basis set are induced by $\Delta E_{gu}/2$.

A comparison of Eq. (26) with Eq. (21) shows that an expression for nonadiabatic transition probability can be written as

$$P_{12} = \exp\left[-\frac{\pi(\Delta E_{gu})^2}{2\hbar\omega\langle g|z|u\rangle E_0}\right]. \qquad (27)$$

This expression is simplified as

$$P_{12} = \exp\left[-\frac{\pi}{2\delta_{t=0}}\right], \qquad (28)$$

where $\delta_{t=0} = \dfrac{\hbar\omega\langle g|z|u\rangle E_0}{(\Delta E_{ug})^2}$, which is the adiabaticity parameter at $t=0$.

Finally, P_{LR}, electron transfer probability from $|L\rangle$ to $|R\rangle$ is given as

$$P_{LR} = 1 - P_{12}. \qquad (29)$$

If $\delta_{t=0} \ll 1$, the adiabaticity condition is fully satisfied and $P_{LR} = 1$, i.e., a complete interwell electron transfer takes place in the first half cycle of the pulse as shown in Fig. 5. As R increases, the adiabaticity parameter increases since energy difference, ΔE_{gu}, becomes small and transition moment, $\langle g|z|u\rangle$, is proportional to R. Effects of nonadiabatic transition, therefore, make a significant contribution and P_{LR} decreases as R increases. This is a semiclassical interpretation of interwell electron transfer of H_2^+.

1.4 Dissociative ionization

Let us now make a quantitative discussion about correlated motions between electrons and nuclei in the laser-induced dissociative ionization of H_2^+. The 3D Hamiltonian presented in this chapter can describe the electronic and nuclear dynamics with equal grounds between them. Since the applied electric field of the laser is parallel to the z direction, the electronic and nuclear wavepackets spread out mainly along the z coordinate, though a small amount of the wavepackets spread out in the ρ direction. To see the correlation behaviors between electronic and

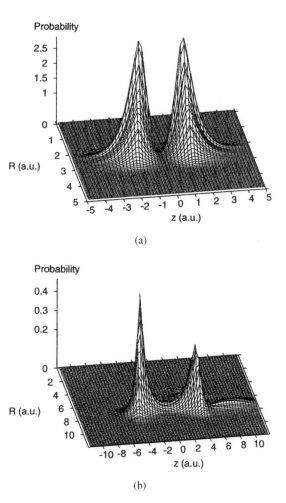

Fig. 7. Snapshots of electronic and nuclear wavepackets in dissociative ionization of H_2^+ at $t = 0$ (a) and $t = 224$ (5.42 fs) (b). These wavepackets are evaluated by Eq. (30). The second pulse is turned on at $t = 0$ just after the transition from $1s\sigma_g$ to $2p\sigma_u$ is completed by the first pulse. The second pulse with $E(t = 224) = -0.0805$ whose intensity is 2.25×10^{14} W/cm^2 is used. Two peaks are bound to the two ion cores in the dissociation channel $H + H^+$ and move along $z = \pm R/2$. (Reprinted with permission from Kawata et al.[1(b)] Copyright (1999) by American Institute of Physics).

nuclear motions of H_2^+ in laser fields, let us evaluate the reduced probability $P(z, R)$ defined as

$$P(z, R) = \int |\phi(\rho, z, R)|^2 \rho d\rho. \tag{30}$$

Figure 7 shows the probability of the dissociative ionization induced by using two pulses: the first pulse is a weak ultrashort pump pulse that excites a hydrogen molecular ion from $1s\sigma_g$ to $2p\sigma_u$, and the second one induces dissociative ionization since the electronic and nuclear wavepackets move on the dissociative potential of

2pσ_u electronic excited state. Parameters of the second pulse are the same as those in Fig. 2. In Fig. 7(a), the probability is drawn as a function of R and z at $t = 0$ when the second pulse is just turned on. The positions of the two peaks correspond to those of the ion cores of H_2^+. In Fig. 7(b), the probability evaluated at $t = 224$ with a laser field with $E(t = 224) = -0.0805$ is shown. An ionizing component starting from nearly 5 au toward larger z can clearly be seen. That is, tunneling ionization (barrier suppression ionization) takes place. When the electric field is minus, $E(t) < 0$, the barrier on the positive z side is low, compared with the barrier on the negative z side. If the electron has enough time to penetrate the barrier before the phase of the electric field changes, tunneling ionization occurs, and if not, multi-photon ionization occurs. The criterion for controlling two types of ionization in intense laser pulses is given by the Keldysh parameter γ as already introduced in Chap. 1.

2. Interelectronic Correlation in a Hydrogen Molecule

Hydrogen molecule is one of the simplest multi-electron systems. In general, in multi-electron systems, electron–electron interaction, called interelectronic correlation, plays an essential role in electron dynamics in intense laser fields. It is well known that in atomic systems, so-called nonsequential multiple ionization that is induced by a breakdown of the single-electron excitation is due to interelectronic correlation.[6] This behavior is observed in momentum distributions of photoelectrons and recoiled ions in He and Ar. Interelectronic correlation exists in molecules as well. The effects of interelectronic correlation in molecules depend on the internuclear distance, which are different from those in atoms. In this section, results of a theoretical study on the interelectronic correlation in H_2 are presented. Two-electron wavepacket dynamics is evaluated by using a multi-dimensional model.

2.1 Hamiltonian of H_2 in the presence of laser fields

Let us consider a hydrogen molecule that is fixed in space and the molecular axis is parallel to the direction of a linearly polarized electric field of an intense laser.[7] The position of the jth electron is expressed in terms of cylindrical coordinates (ρ_j, z_j, ϕ_j) in the same way as that in H_2^+. The origin of the z-axis is set at the center of the chemical bond with internuclear distance R. The Hamiltonian of H_2 in laser fields, $H(t)$, is written as the sum of one-electron Hamiltonian with dipole interaction, $H_j^0(t)$, and two-electron Hamiltonian, H_{12}^0, as

$$H(t) = \sum_{j=1}^{2} H_j^0(t) + H_{12}^0. \qquad (31)$$

Here, one-electron Hamiltonian $H_j^0(t)$ in the laser field is given in atomic units as

$$H_j^0(t) = -\frac{1}{2}\left(\frac{\partial^2}{\partial \rho_j^2} + \frac{1}{\rho}\frac{\partial}{\partial \rho_j} + \frac{\partial^2}{\partial z_j^2}\right) + V(\rho_j, z_j) + z_j E(t), \qquad (32)$$

where $E(t)$ is an electric field of laser and $V(\rho_j, z_j)$, the Coulomb interaction between the jth electron and the nuclei, is given as

$$V(\rho_j, z_j) = -\frac{1}{\sqrt{\rho^2 + \left(z - \frac{R}{2}\right)^2}} - \frac{1}{\sqrt{\rho^2 + \left(z + \frac{R}{2}\right)^2}}. \qquad (33)$$

The kinetic energy of the two electrons rotating around the z-axis, $-\sum_{j=1}^{2}\frac{1}{2\rho_j^2}\frac{\partial^2}{\partial \varphi_j^2}$, is included in H_{12}^0.

The two-electron Hamiltonian is given as

$$H_{12}^0 = -\sum_{j=1}^{2}\frac{1}{2\rho_j^2}\frac{\partial^2}{\partial \varphi_j^2} + V_{12}, \qquad (34)$$

where V_{12}, electron–electron repulsion energy, is expressed as

$$V_{12} = \frac{1}{\sqrt{\rho_1^2 + \rho_2^2 - 2\rho_1\rho_2\cos(\varphi_1 - \varphi_2) + (z_1 - z_2)^2}}. \qquad (35)$$

The two-electron Hamiltonian is written after transforming the two variables φ_1 and φ_2 into the relative angle $\phi = \varphi_1 - \varphi_2$ and averaging $\chi = \frac{1}{2}(\varphi_1 + \varphi_2)$ as

$$H_{12}^0 = -\frac{1}{2}\left(\frac{1}{\rho_1^2} + \frac{1}{\rho_2^2}\right)\left(\frac{\partial^2}{\partial \phi^2} + \frac{\partial^2}{4\partial \chi^2}\right) - \frac{1}{2}\left(\frac{1}{\rho_1^2} - \frac{1}{\rho_2^2}\right)\frac{\partial^2}{\partial \phi \partial \chi}$$
$$+ \frac{1}{\sqrt{\rho_1^2 + \rho_2^2 - 2\rho_1\rho_2\cos\phi + (z_1 - z_2)^2}}. \qquad (36)$$

The z-axis component of the total angular momentum whose operator is given as $-i\frac{\partial}{\partial \chi}$ is conserved since the electric field is applied parallel to the axis. The two-electron Hamiltonian is reduced to the effective two-electron Hamiltonian in which $-i\frac{\partial}{\partial \chi}$ is replaced by a quantum number $\ell = 0, \pm 1, \pm 2, \ldots$ as

$$H_{12}^0(\ell) = -\frac{1}{2}\left(\frac{1}{\rho_1^2} + \frac{1}{\rho_2^2}\right)\left(\frac{\partial^2}{\partial \phi^2} - \frac{\ell^2}{4}\right) - \frac{i\ell}{2}\left(\frac{1}{\rho_1^2} - \frac{1}{\rho_2^2}\right)\frac{\partial}{\partial \phi}$$
$$+ \frac{1}{\sqrt{\rho_1^2 + \rho_2^2 - 2\rho_1\rho_2\cos\phi + (z_1 - z_2)^2}}. \qquad (37)$$

The total Hamiltonian is finally expressed as the sum of the one-electron Hamiltonian and the effective two-electron Hamiltonian as

$$H(t) = \sum_{j=1}^{2}\left(-\frac{1}{2}\left(\frac{\partial^2}{\partial\rho_j^2} + \frac{1}{\rho}\frac{\partial}{\partial\rho_j} + \frac{\partial^2}{\partial z_j^2} + \frac{1}{\rho_j^2}\left(\frac{\partial^2}{\partial\phi^2} - \frac{\ell^2}{4}\right)\right) + V(\rho_j, z_j) + z_j E(t)\right)$$
$$-\frac{i\ell}{2}\left(\frac{1}{\rho_1^2} - \frac{1}{\rho_2^2}\right)\frac{\partial}{\partial\phi} + \frac{1}{\sqrt{\rho_1^2 + \rho_2^2 - 2\rho_1\rho_2\cos\phi + (z_1 - z_2)^2}}. \quad (38)$$

2.2 Electronic wavepacket dynamics of H_2

A hydrogen molecule has now five variables ($\rho_1, \rho_2, z_1, z_2, \phi$). To see the dynamic behaviors of the electronic wavepackets in a laser field, reduced density $\bar{P}(z_1, z_2)$ is evaluated. $\bar{P}(z_1, z_2)$ is defined as

$$\bar{P}(z_1, z_2) = \int_0^\infty d\rho_1 \int_0^\infty d\rho_2 \int_0^{2\pi} d\phi |\Phi(t)|^2. \quad (39)$$

Here, $\Phi(t)$ is the solution of the time-dependent Schrödinger equation of H_2. The reduced density is appropriate for describing the electronic dynamics since electron motions are mainly along the z direction, i.e., the polarization direction of the applied laser field. The electric field of the laser is set to $E(t) = f(t)\sin(\omega t)$, where the envelope function $f(t)$ is linear with respect to t and reaches its maximum f_0 after one cycle. The field parameters used are $f_0 = 0.12 E_h/ea_0$ ($I = 5.04 \times 10^{14}$ W/cm^2) and $\omega = 0.06 E_h/\hbar$ (wavelength $\lambda = 760$ nm).

The time-dependent Schrodinger equation is solved by using the alternating direction implicit method described in Chap. 3. The time step is taken to be $\Delta t = 0.05\hbar/E_h$. The grid end points are set to $\rho_{max} = 5a_0$ and $z_{max} = -z_{min} = 12a_0$.

Figure 8 shows snapshots of the reduced density of H_2 in laser fields. The internuclear distance is fixed at $R = 4a_0$, i.e., at the equilibrium nuclear configuration in the ground electronic state.[7] In Fig. 8(a), $\bar{P}(z_1, z_2)$ for the ground sate $X^1\Sigma_g^+$ is plotted. This clearly demonstrates the covalent character of H_2 in the ground state: the densities are located around two positions, $(z_1, z_2) = (2a_0, -2a_0)$ and $(-2a_0, 2a_0)$, i.e., the electronic wavefunction is expressed as $\Phi(t = 0) \approx a(1)b(2) + b(1)a(2)$, where a and b are the $1s$ atomic orbitals on the right and left protons, respectively, and 1 and 2 denote the coordinates of the two electrons.

Figure 8(d) shows two types of the ionization mechanism: one is direct ionization from the ground electronic configuration whose path is indicated by dotted arrows. The other is indirect ionization *via* an intermediate electronic configuration with an ionic character whose path is indicated by broken arrows. The latter is a nonsequential double ionization that is brought about by an interelectronic correlation

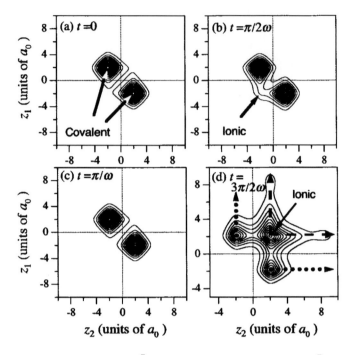

Fig. 8. Snapshots of reduced density $\bar{P}(z_1, z_2)$ of H_2 at $R = 4.0a_0$. (a) $\bar{P}(z_1, z_2)$ at $t = 0$, (b) $t = \pi/2\omega = 26.2\hbar/E_h$, (c) $t = \pi/\omega$, and (d) $t = 3\pi/2\omega$ with $\omega = 0.06 E_h$. Ionic component $H^- H^+$ appears around $z_1 = z_2 = -R/2$ as the electric field approaches the local minima as shown in (b). In (d), on the other hand, ionic component $H^+ H^-$ appears. This indicates that the two electrons adiabatically follow the electric field. Broken arrows indicate the ionization path *via* the ionic state because of a strong interelectronic correlation. Dotted arrows indicate direct ionization from the ground state. (Reprinted with permission from Haruyama *et al.*[7] Copyright (2002) by the American Physical Society).

in an intense laser field. Nonsequential double ionization of H_2 was experimentally observed in the $10^{13} - 10^{14}$ W/cm^2 intensity range using 10-fs laser pulses.[8]

Figure 9 shows $\bar{P}(z_1, z_2)$ of H_2 at $R = 6a_0$. The reduced density in the absence of the electric field indicates a covalent bond of H_2, as shown in Fig. 9(a), while the ionic component $|\langle \Phi | H^+ H^- \rangle|^2$ is around 0.016. Onset of rapid ionization from the ionic state created by the pulse laser can be seen in Figs. 9(b) and 9(c). This is because of no overlap between the covalent and ionic states at long the internuclear distance R.

Double ionizations *via* an electronic intermediate state with an ionic character are expected to bring about dissociations with charge-asymmetric fragments. For example, asymmetric pathways in a nitrogen molecule have been observed[8]: the ionization is from (1,2) to (1,3) and a symmetric pathway is the ionization from (1,2) to (2,2), where (n, m) stands for the $n + m$ charged state leading to the dissociation channel $N_2^{(n+m)+} \rightarrow N^{n+} + N^{m+}$.

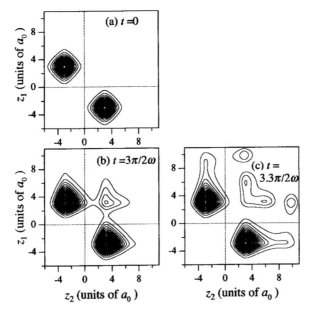

Fig. 9. Snapshots of reduced density of H_2 at $R = 6.0a_0$. (a) $t = 0$, (b) $t = 3\pi/2\omega$, (c) $t = 3.3\pi/2\omega$, and (d) $t = 3\pi/2\omega$. Laser parameters are the same as those in Fig. 8. (Reprinted with permission from Haruyama et al.[7] Copyright (2002) by the American Physical Society).

3. Reaction Dynamics of Carbon Dioxide

In this section, experimental and theoretical studies of reaction dynamics of CO_2 in intense infrared laser pulses are presented. One optical cycle-averaged potential energy surface is introduced to analyze the nuclear dynamics.

3.1 Doorway states for Coulomb explosions

Results of experiments on Coulomb explosions of CO_2 in intense infrared laser pulses have been reported in detail.[10,11] In intense fields, CO_2 is successively ionized at large internuclear distances, and multiply charged cations undergo Coulomb explosion. The geometrical structures of the doorway states leading to Coulomb explosion are experimentally determined by using sophisticated methods of detection such as measurements of distribution patterns in a covariance map of fragments ions, momentum vector distribution with mass-resolved momentum imaging, and coincidence momentum imaging. For example, for CO_2^{3+} irradiated by a 100-fs pulse with 1.1 PW/cm^2 ($\lambda = 795$ nm), the C–O bond length is stretched to 1.7 Å and the mean bending angle is about 20° as measured from the linear geometry.[11]

Let us now qualitatively explain the observed geometrical structure using both the time-dependent adiabatic state approach and the concept of electron transfer between nuclei in an intense laser field.[12] The precursor configuration leading to CO_2^{3+} is assigned to CO_2^{2+} among the neutral CO_2, monocation CO_2^+ and dication CO_2^{2+} from calculations of time-dependent adiabatic potential energies. Carbon dioxide keeps its stable linear structure in intense laser fields with about 1 PW/cm^2 and the geometrical structures of the lowest two electronic states, $^2B_2$ and $^2A_2$, in bent CO_2^+ keep their stable structures in these intense laser fields.

For CO_2^{2+}, there exist three low electronic states: the ground triplet state $^3B_1$ ($^3\Sigma_g$ at $D_{\infty h}$ linear geometry) and the two singlet states $^1A_1$ and $^1B_1$ ($^1\Delta_g$). The energy difference between the $^1A_1$ and the $^3B_1$ states in a field-free condition is about 1.5 eV, which is nearly equal to the energy of one photon at the equilibrium geometry. It is theoretically shown that the potential energy surfaces of these three states are expressed almost in the same form. These above results indicate that these three states are equally populated from CO_2^+.

The calculated time-dependent adiabatic potential energy surfaces of $^1A_1$ CO_2^{2+} are shown in Fig. 10. A linearly polarized electric field $E(t)$ is applied along the molecular axis. The potential energy surfaces are calculated by a simple CI method with the MCSCF orbitals obtained at zero field. Here, 2D potential energies are drawn as a function of the C–O internuclear bond distance and the O–C–O bending angle at several field strengths: (a) $E(t) = 0$, (b) $E(t) = 0.1 E_h/ea_0$, and (c) $E(t) = 0.2 E_h/ea_0$. With increase in field strength, a considerable deformation of the geometrical structure occurs: the C–O bond stretches from the linear structure to a bent structure accompanied by geometrical structure changes from linear to a bent structure of CO_2^{2+}.

The streams of arrows shown in Fig. 10(c) denote a typical path of the geometrical changes. Results of Mulliken population analysis show that two positive charges in CO^{2+} are nearly equally distributed among the three atoms. Three electronic configurations, OC^+O^+, O^+C^+O, and $O^-C^+O^{2+}$, are possible. Among them, $O^-C^+O^{2+}$ is an ionic configuration favorable for interelectronic transfer from OC^+O^+ to CO_2^{3+}, the same as that of H^-H^+ or H^+H^- inducing double electron ionizations in H_2. The analysis of time-dependent potential energies at an instantaneous time shows that field-induced bond stretching is accompanied by a large amplitude bending motion. This is responsible for the observed geometrical structure of precursor CO_2^{3+} leading to Coulomb explosions.

3.2 Cycle-averaged potential energy surface and simultaneous two-bond breaking

Unlike electronic dynamics, nuclear dynamics of CO_2 in an intense alternating electric field of a laser is not governed by instantaneous time-dependent potential $V(\mathbf{R}, t)$. Here, $\mathbf{R}$ denotes the internal coordinates. This is true for other molecules. This is because ω (laser frequency) $\gg \omega_{vib}$ (molecular vibration), and each atom

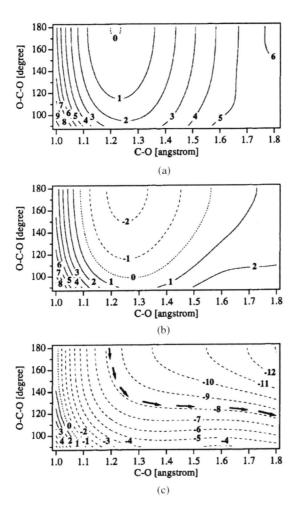

Fig. 10. 2D potential energy surfaces of the lowest singlet state (connected with $^1A_1$ at zero field) of CO^{2+} at (a) $E(t) = 0$, (b) at $E(t) = 0.1E_h/ea_0$, and (c) at $E(t) = 0.2E_h/ea_0$. (Reprinted with permission from Kono et al.[12] Copyright (2001) by American Chemical Society).

in CO_2 cannot adjust to changes in the electric field $E(t)$. Since pulse envelope function $f(t)$ does not change in an optical cycle of a laser in ordinary experiments except for a few cycle pulse experiments, i.e., $\omega \gg df(t)/dt/f(t)$, nuclear dynamics can be analyzed based on a cycle-averaged time-dependent potential $\bar{V}(R, t)$. The potential $\bar{V}(R, t)$ can be expanded in terms of the nth power of the electric field, $E^n(t)$, averaged over one optical cycle[13]

$$\frac{\omega}{2\pi} \int_{t-\frac{\pi}{\omega}}^{t+\frac{\pi}{\omega}} E^n(t) dt \quad \text{as}$$

$$\bar{V}(R, t) = \bar{V}(R, t=0) - \frac{1}{4}\alpha(R) f^2(t) - \frac{3}{32}\gamma(R) f^4(t) + \cdots$$

Here, $\alpha(R)$ and $\gamma(R)$ are the polarizability and hyperpolarizability terms, respectively. The odd-order terms with respect to $E(t)$, such as the dipole interaction term, vanish. The effective potential consists of even-order terms of the pulse envelope function $f(t)$. For comparison between the instantaneous and effective potentials, the potential energy surfaces of the lowest adiabatic state of CO_2 are drawn as a function of the two C–O bond distances R_1 and R_2 in Fig. 11.

The features of the optical cycle-averaged potential are that the energy is reduced and the potential becomes symmetric rather than antisymmetric with respect to exchange between R_1 and R_2. Concerning the reduction in energy, the dissociation energy for the C–O bond is about 7 eV in a zero field, while the energy is reduced

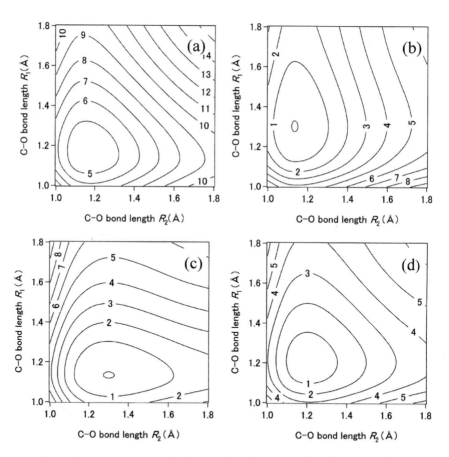

Fig. 11. Potential energy surfaces of the lowest adiabatic state of CO_2. The contour lines are plotted at intervals of 1 eV. (a) Potential in the absence of the electric field of a laser, (b) adiabatic potential $V(R_1, R_2, t)$ at $E(t) = 0.1\ E_h/ea_0$, (c) $V(R_1, R_2, t)$ at $E(t) = -0.1\ E_h/ea_0$, and (d) $\bar{V}(R_1, R_2, t)$, one optical cycle-averaged effective potential. (Reprinted with permission from Sato et al.[13] Copyright (2003) by American Chemical Society.)

to 4 eV at $E(t) = 0.1 E_h/ea_0$. This reduction mainly originates from the polarization energy, the second term in the effective potential $\bar{V}(\mathbf{R}, t)$. It should be noted in Fig. 11 that the equilibrium geometry in $\bar{V}(R_1, R_2, t)$ is almost identical to that in a zero field. The equilibrium internuclear distance of the C–O bond at $f(t) = 0.14 E_h/ea_0$ is calculated to be 1.2 Å.

Let us now consider dissociation dynamics on the lowest adiabatic potential energy surface $\bar{V}(R_1, R_2, t)$ of linear CO_2 in an intense field. There are two pathways to breaking C–O bonds in the linear form, one-bond breaking and two-bond breaking. The dissociation energy in a zero field is ca. 11 eV for the two-bond breaking, while it is ca. 1 eV for the one-bond breaking. Therefore, from the viewpoint of dissociation energy, one-bond breaking is preferred to two-bond breaking in a zero field.

Figure 12 shows the nuclear wavepackets of CO_2^{2+} in the lowest state in an intense field with peak field strength of $0.19 E_h/ea_0 (1.3 \times 10^{15}$ W/cm^2), pulse length of 194 fs, and $\lambda = 795$ nm. It is assumed that the initial wavefunction is prepared in the lowest ground state of CO_2^{2+} at $t = 0$. In Fig. 12(a), the square of the initial wavefunction is indicated by pink contour circles around $R_1 = R_2 = 1.2$ Å. At $t = 24.7$ fs, the wavepacket propagates toward the one-bond breaking path from the initial Franck–Condon position as shown by the blue deformed contour circles. The instantaneous potential at $E(t = 24.7$ fs$) = -0.029 E_h/ea_0$ is shown by green

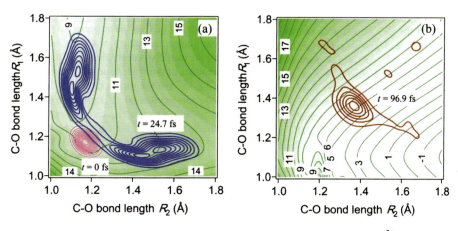

Fig. 12. Nuclear wavepacket dynamics on the lowest adiabatic potential of CO_2^{2+} in an intense field. (a) A pink contour circle around $R_1 = R_2 = 1.2$ Å is the square of the initial wavefunction. The instantaneous potential at $E(t = 24.7$ fs$) = -0.029 E_h/ea_0$ is shown by a green contour line. The wavepacket denoted by blue deformed contour circles propagates toward the one-bond breaking path from the initial Franck–Condon position. (b) The square of the wavepacket at $E(t = 96.9$ fs$) = 0.19 E_h/ea_0$ is denoted by red contour circles. The wavepacket propagates toward the two-bond dissociation as well as the one-bond dissociation. (Reprinted with permission from Sato et al.[13] Copyright (2003) by American Physical Society).

contour lines. With passage of time, the main wavepacket denoted by blue color does not reach the region of the one-bond dissociation fragments but returns to the vicinity $R_1 = R_2 = 1.2$ Å at $t = 45$ fs. In Fig. 12(a), the square of the wavepacket at $E(t = 96.9\,\text{fs}) = 0.19 E_\text{h}/ea_0$ is plotted. This figure shows that the symmetric stretching mode is highly excited and that the wavepacket propagates toward the two-bond dissociation as well as the one-bond dissociation.

4. Benzene

Mechanisms of Coulomb explosion in diatomic molecules or small polyatomic molecules such as CO_2, C_2H_2, and C_2H_4 have been investigated intensively both experimentally and theoretically.[14] Multi-charged atomic and molecular ion fragments are detected by time-of-flight spectroscopy combined with the covariance mapping technique as described in Chap. 2.[15] Coulomb explosion in small molecules takes place at a critical bond distance R_c that is about two-times longer than the equilibrium bond distance. Coulomb explosion in some polyatomic molecules such as furan (C_4H_4O), pyrrole (C_4H_4NH), pyridine (C_5H_5N), and pyrazine ($C_4H_4N_2$) that belong to the family of hetero and cyclic aromatic molecules takes place at a critical geometry with an extended bond distance as well.[16]

Mass spectra from the ionization and fragmentation of benzenes in intense laser beams have been observed with various laser parameter conditions.[17] Direct proof of Coulomb explosions in benzene can be obtained by measuring the large kinetic energies of fragment C^{q+} ions.

Figure 13 shows ion intensities and average kinetic energies of multi-charged carbon atoms at $2.9 \times 10^{16} \sim 1.8 \times 10^{17}$ W cm^{-2} intensities of fs laser beams with a central wavelength of 800 nm.[18] This figure indicates that Coulomb explosion of the fragments from the dissociation of neutral benzene cannot explain the present large kinetic energies of the C^{q+} ions, but the large kinetic energies can be explained if the Coulomb explosion takes place from $C_6H_6^{n+}$. The results of the experiment also indicate that Coulomb explosions in benzene take place at a critical geometry with slight expansion.

By using a tandem time-of-flight mass (TOF–MS) spectrometer, Itakura et al. investigated the ionization and dissociation mechanisms of benzenes.[19] The TOF–MS spectrometer consisted of two stages: In the first stage, benzene cations were spatially focused and in the second stage, the mass-selected cations were ionized and fragmented by intense laser pulses. In this experiment, intense femtosecond laser pulses with wavelengths of $\lambda = 395$ nm and $\lambda = 790$ nm and intensity of $\sim 2 \times 10^{16}$ and pulse duration of ~ 50 fs were used. The values of the Keldysh parameter, γ, were $\gamma = 0.24$ and $\gamma = 0.12$ for $\lambda = 395$ nm and $\lambda = 790$ nm, respectively. This

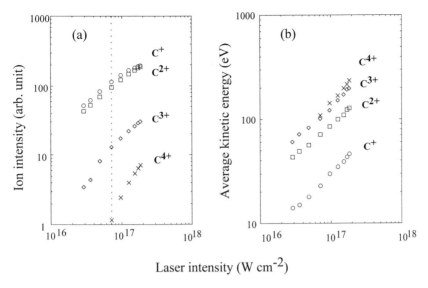

Fig. 13. (a) Dependence of carbon ion (C^{q+})($q = 1 \sim 4$) intensities on laser intensity from $2.9 \times 10^{16} \sim 1.8 \times 10^{17}$ W cm^{-2} for benzene. The slope at an increasing rate of C^{3+} ions is slowed down by the appearance of C^{4+} ions at 7.1×10^{16} W cm^{-2}, which is denoted by the dotted line. (b) Dependence of the average kinetic energies of carbon ions on laser intensities. The kinetic energies increased with increase in the laser intensity. (Reprinted with permission from Shimizu *et al.*[18] Copyright (2001) by Elsevier).

means that tunneling ionization is expected for both cases. Furthermore, the tendency of tunneling ionization in the latter case is stronger than that in the former case.

Figure 14 shows the TOF mass spectra obtained when neutral benzenes were irradiated by intense laser pulses with wavelengths of (a) 395 nm and (b) 790 nm.[19] The main product ions were benzene cations at both laser wavelengths, but for (a) the yields of other product ions, $C_6H_5^+$, $C_5H_3^+$, $C_4H_i^+$ ($i = 2 - 4$), $C_3H_i^+$ ($i = 1 - 3$), $C_2H_2^+$ and $C_2H_3^+$ were obtained.

Figure 15 shows the TOF mass spectra obtained when mass-selected benzene cations were irradiated by intense laser pulses with wavelengths of (a) 395 nm and (b) 790 nm.[19] For (a), $C_4H_i^+$ ($i = 2 - 4$) fragment ions were obtained as the major products, while $C_3H_i^+$ fragment ions were obtained with much weaker intensities. For (b), on the other hand, benzene dications appeared with strong peak intensity, while there were no $C_3H_i^+$ fragment ions.

It can be seen from a comparison of Figs. 14 and 15 that for both laser wavelengths, the major products obtained when starting from benzene cations were the same as those having the largest yields next to the benzene cations when starting from neutral benzene molecules. This suggests that the ionization and fragmentation dynamics starting from neutral benzene molecules are governed by processes common to those starting from benzene cations.

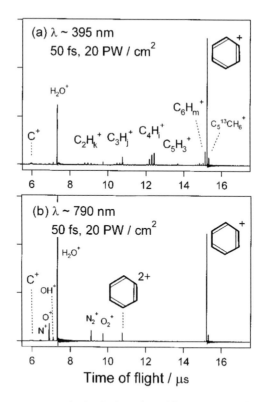

Fig. 14. The TOF mass spectra obtained when neutral benzenes were irradiated by intense laser pulses with wavelengths of (a) 395 nm and (b) 790 nm. (Reprinted with permission from Itakura et al.[19] Copyright (2000) by American Institute of Physics).

For $\lambda = 395$ nm (3.14 eV) excitation, benzene cations are populated in the electronic ground $\tilde{X}^2 E_{1g}$ state by three-photon excitation as shown in the schematic diagram of the ionization and fragment processes in Fig. 16. Then, they are resonantly excited to the $\tilde{C}^2 A_{2u}$ state by one-photon absorption of $\lambda = 395$ nm. Photon-dressed potential energy surfaces are expected to be formed between $\tilde{C}$ and $\tilde{X}$ states since the $\tilde{C}-\tilde{X}$ transition is a strong dipole-allowed transition. Furthermore, the $\tilde{C}$ state of benzene cations undergoes ultrafast nonradiative transition to lower electronic states, for example, the $\tilde{X}$ state with a vibrationally highly excited manifold. Further irradiation of intense laser beams induces the formation of complex fragment ions, $C_4H_i^+$ and $C_3H_i^+$.

For $\lambda = 790$ nm excitation, benzene dications and trications were produced, while benzene cations and dications were produced when starting from neutral benzene molecules. The efficient ionization indicates that there is no strong coupling among specific electronic states on either of the neutral, cation, dication species and that nonresonant multiple field-induced tunnel ionization dominates.

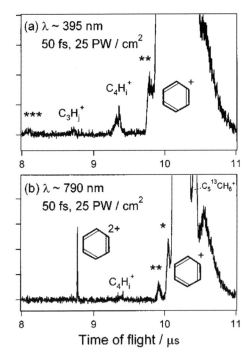

Fig. 15. The TOF mass spectra obtained when mass-selected benzene cations were irradiated by intense laser pulses with wavelengths of (a) 395 nm and (b) 790 nm. (Reprinted with permission from Itakura et al.[19] Copyright (2000) by American Institute of Physics).

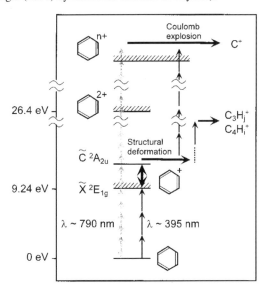

Fig. 16. Schematic diagram of the ionization and fragmentation processes of benzene in intense laser fields. Shorter and longer vertical arrows represent photons with wavelengths of 790 nm and 395 nm, respectively. (Reprinted with permission from Itakura et al.[19] Copyright (2000) by American Institute of Physics).

5. Fullerene

Electronic and vibrational dynamics of fullerenes in an intense laser field is a research subject of great interest. This is because fullerenes belong to a large finite system and it is expected that the dynamics can be explained in terms of mechanisms that have no so far been elucidated. On the other hand, a representative of fullerenes, Buckminster fullerene C_{60}, is highly symmetric and belongs to the I_h irreducible representation of the point group; its geometrical structure consists of 12 pentagons and 20 hexagons. Therefore, it is possible to construct a model for the dynamics of C_{60} in analyzing the dynamics. There have been a considerable number of studies on electronic and nuclear dynamics of C_{60} under various kinds of pulse excitation conditions.[20]

Figure 17 shows typical mass spectra obtained from C_{60} by applying laser pulses of $\lambda = 765$ nm with pulse duration of 25 fs and with pulse duration of 5 ps.[20(g)]

The spectra were recorded at roughly equal laser fluence. The intensities corresponding to the top and bottom spectra are $\sim 3.2 \times 10^{12}$ W cm^{-2} and 1×10^{15} W cm^{-2}, respectively, which correspond to Keldysh parameters of 4.4 and 0.25, respectively. That is, these belong to multi-photon and tunneling regimes, respectively. For 5-ps pulses, a normal bimodal mass spectrum that has been observed in other photofragmentation experiments appears. Many singly charged fragments are produced together with singly charged parent ions. The main peaks in the mass spectrum are due to dissociation products, C_{60-2n}^+. For 25-fs pulses, on the other hand, a series of multiply charged C_{60}^{z+} ions ($z = 1, 2, \ldots, 5$) are observed together with their large

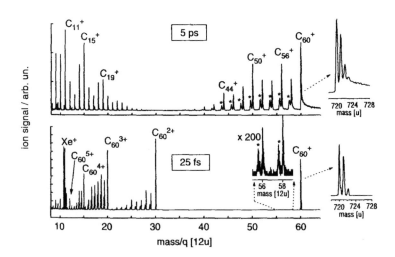

Fig. 17. Typical mass spectra obtained from C_{60} by applying laser pulses of $\lambda = 765$ nm with pulse duration of 25 fs (bottom) and 5 ps (top). (Reprinted with permission from Hertel *et al.*[20(g)] Copyright (2005) by Institute of Physics).

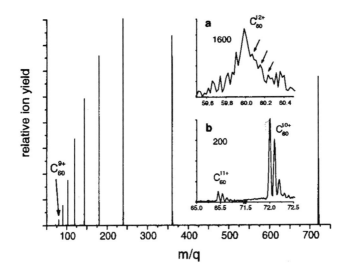

Fig. 18. Production of highly charged C_{60}^{z+} cations, $z \leq 12$, by using 70-fs laser pulses with wavelength of 1800 nm and intensity of 10^{15} W cm^{-2}. The insets show the low m/q regions corresponding to C_{60}^{z+}, $z = 10$–12, at increased magnification. Peak heights for $z < 9$ do not reflect mass abundance due to saturation. (Reprinted with permission from Bhardwaj et al.[22] Copyright (2003) by the American Physical Society).

fragments by C_2 evaporation with their small probability. These multiply charged parent ions are not seen for 5-ps pulse excitation. In both cases, the fragment ions are in vibrationally hot states just after the fragmentation takes place and then they dissociate on a μs-to-ms time scale within which the system is in an equilibrium condition. The fragment peaks denoted by asterisks are due to a delayed ionization.[21]

How many positive charges can exist in the stable C_{60}^{z+}? Figure 18 shows the mass spectrum obtained by irradiation of infrared pulses with $\lambda = 1800$ nm, pulse duration of 70 fs, and intensity of 10^{15} W cm^{-2}. It can be seen that highly charged C_{60}^{z+} cations, $z \leq 12$, are produced. C_{60}^{12+} can survive as intact parent ions that are stable at least over the 0.5 μs.[22]

The stability of highly charged fullerene C_{60}^{12+} has been theoretically explained by using the density functional theory with the B3LYP functional for exchange and correlation.[23] Here, the optimal geometries and electronic properties of positively charged C_{60}^{z+} fullerenes ($z = 0 - 14$) were calculated. It is concluded from the calculation that the Coulomb stability limit corresponds to $z = 14$.

Several models for explaining such stability and fragmentation of highly charged C_{60} have been proposed. These models include (a) a transient thermal electron emission model in which the electrons can be considered hot and the vibrational degrees of freedom can be considered cold[24]; (b) a sequential nonadiabatic excitation model in which the following three steps are considered: (i) nonadiabatic, population transfer from the ground electronic state to the excited-state manifold *via* a doorway, charged transfer state, (ii) exponential enhancement of this transition by collective dynamic polarization of all the electrons, and (iii) sequential energy deposition in both neutral molecules and resulting ions[25]; and (c) multi-photon excitation of giant plasmon.[26]

Let us now discuss how much vibrational energy is acquired by C_{60} or C_{60}^{+z} under intense ultrashort infrared pulse conditions. For this purpose, a quantum mechanical method for the evaluation of nuclear wavepacket motions on a time-dependent and adiabatic potential surface given in Sec. 3.2 has been adopted.[27] The possibility of vibrational excitations of C_{60} and C_{60}^{+z} due to field-induced dipole forces has been pointed out.[22] For simplicity, it is assumed that the impulsive excitations are separated from vertical ionizations at the I_h structure of C_{60}. The ground vibrational state of C_{60} or C_{60}^{+z} is chosen as each initial state. Here, impulsive vibrational Raman processes by a Gaussian pulse of $\lambda = 1800$ nm, pulse width of 70 fs, and different peak intensities I_{peak} of 1.0, 0.9, 0.8, and 0.7 $\times$ 10^{15} W cm^{-2} are considered. The nonadiabatic transitions to excited electronic states, i.e., higher time-dependent adiabatic states of C_{60} or C_{60}^{+z}, are omitted because their probabilities are negligible for 1800 nm excitation. The nuclear motion of $h_g(1)$ mode upon excitation is along the direction of the polarization of laser pulses. The lowest time-dependent (instantaneous) adiabatic potential $V_1(R_{hg}, t)$ is obtained by solving the time-independent Schrödinger equation at different static field strengths. It has been found that the acquired energy is nearly proportional to $I_{peak}^{2.5}$, indicating that the order of the nonlinear processes in the intensity regime is slightly higher that of the ordinal, impulsive Raman excitation process.[27]

The vibrational wavepacket $X_1(R_{hg}, t)$ on the instantaneous potential energy $V_1(R_{hg}, t)$ satisfies the time-dependent Schrödinger equation:

$$-i\hbar \frac{\partial}{\partial t} X_1(R_{hg}, t) = \left[-\frac{\hbar^2}{2\mu} \frac{\partial^2}{\partial R_{hg}^2} + V_1(R_{hg}, t) \right] X_1(R_{hg}, t).$$

Here, μ is the reduced mass of the $h_g(1)$ mode.

Figure 19 shows the propagation of wavepacket $|X_1(R_{hg}, t)|$ along the $h_g(1)$ mode.[27] The geometrical structures of C_{60} at different values of R_{hg} are illustrated above the panel. The positive value of R_{hg} indicates a prolate form of C_{60} and the negative value indicates an oblate form. The line with up and down arrows by the prolate form indicates the polarization direction of light. The absolute value of the wavepacket $|X_1(R_{hg}, t)|$ is denoted by solid contour lines. The cycle-averaged potential $\overline{V}_1(R_{hg}, t)$ is denoted by broken contour lines at intervals of 5 eV. The potential minimum is set to zero irrespective of time. The dark shade indicates the lower energy area of $\overline{V}_1(R_{hg}, t)$. Through interaction with the applied pulse, the $h_g(1)$ vibrational mode is excited to have a large-amplitude oscillation; the acquired energy due to impulsive Raman excitation is 22 eV.

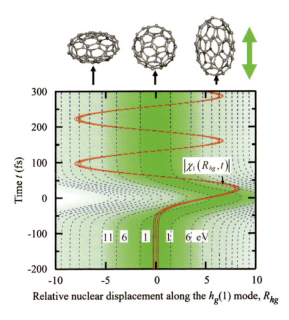

Fig. 19. Propagation of the vibrational wavepacket $|X_1(R_{hg}, t)|$ along the $h_g(1)$ vibrational coordinate R_{hg} of C_{60} when an ultrashort intense Gaussian pulse centered at time $t = 0$ fs ($\lambda = 1800$ nm) is applied. The full width at half maximum of the pulse is 70 fs and the peak light intensity I_{peak} is 10^{15} W cm^{-2}. Broken contour lines denote the cycle-averaged potential $\overline{V}_1(R_{hg}, t)$ at intervals of 5 eV. (Reprinted with permission from Sahnoun et al.[27(a)] Copyright (2006) by American Institute of Physics).

It has also been found that the $h_g(1)$-like modes for high charge states are as robust as in the neutral case, regardless of whether or not an electric field is applied.[27] The vibrational frequency of the $h_g(1)$-like mode of C_{60}^{+12} is reduced only by about 20%. The potential minimum position of the $h_g(1)$-like mode is almost the same as that for neutral fullerene. The acquired energy in the $h_g(1)$-like mode is as large as about 15 eV, which is larger than the fragment appearance energy of C_{60}^{+12}. This indicates that even positive charge cations of fullerenes with large-amplitude vibrational motion along the $h_g(1)$-like mode can survive for a period of μs.[27]

Fullerenes are attractive molecules from the viewpoint of applied science as well as pure science. They have fascinating properties such as superconductivity[28] and a large optical nonlinearity.[29] This originates from their molecular sizes of the large finite system with π electrons that may delocalize on their surfaces. For example, the optical polarizability of Buckminster fullerenes C_{60} at the fundamental wavelength of an Nd:YAG laser (1.064 μm) has been measured to be $\alpha = 79$ Å^3.[29] This high optical nonlinearlity of C_{60} suggests that fullerenes are also possible candidates for a material of higher-harmonic generation (HHG) for producing coherent extreme ultraviolet (XUV) radiation.[20(c),30]

6. Ejection of Triatomic Hydrogen Molecular Ions from Hydrocarbons

Fragment molecular and atomic ions produced by Coulomb explosion of polyatomic molecules are due to the breaking of chemical bonds as shown in the preceding section. Interesting chemical reaction dynamics of polyatomic molecules in Coulomb explosion have been reported by Yamanouchi's group. They found triatomic H_3^+ and diatomic H_2^+ hydrogen molecular ions in the TOF mass spectra of fragment ions of methanol from CH_3OH after irradiation of infrared (800 nm wavelength) laser pulses of 86 fs with intensity $\sim 10^{14}$ W/cm^{-2}. To identify the process of the ejection of H_3^+ from methanol in an intense laser field, similar experiments were performed for a deuterium species, CD_3OH.

Figures 20(a) and 20(b) show TOF mass spectra of fragment ions in the mass range of $m = 1 - 6$ u/z, which are produced from CH_3OH and CD_3OH, respectively.[31] The ejected species consist of triatomic hydrogen molecular ions H_3^+ and HD_2^+ in addition to diatomic hydrogen molecular ions H_2^+ and D_2^+. These species exhibit a triplet pattern: two large side peaks appearing on both sides of a small central peak. The side peaks are assigned to the fragment ions ejected along the forward and backward directions of the TOF axis.

The ejection process of triatomic hydrogen molecular ions is thought to consist of two steps: formation of doubly charged methanol ions, CH_3OH^{2+} (CD_3OH^{2+}), and chemical reactions (direct dissociation, H/D exchange) occurring in the doubly charged parent ions.

The doubly charged parent ions are formed within the pulse duration of 86 fs, and chemical reactions take place on the time scale of 1–10 ps, which is comparable to rotation of the parent ion. The first step was assigned from the experimental

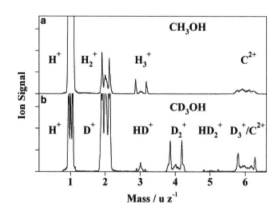

Fig. 20. Time-of-flight (TOF) mass spectra of fragment ions produced from (a) CH_3OH and (b) CD_3OH. (Reprinted with permission from Furukawa et al.[31] Copyright (2005) by Elsevier).

results obtained by using coincident momentum imaging (CMI). The second step was assigned by analysis of mass-resolved momentum imaging (MRMI) maps of ejected molecular ions.

Triatomic hydrogen molecular ions are also ejected from other hydrocarbon molecules such as ethanol, 1-propanol, acetone, acetaldehyde, methane, ethane, ethylene, allene, 1,3-butadiene, and cyclohexane. Therefore, the production of triatomic hydrogen molecular ions from hydrocarbons in intense fields is not peculiar event but a general event, though its detailed mechanisms have not been elucidated.[32]

In summary, first fundamental concepts in theoretical treatments of electronic and nuclear dynamics of molecules in intense laser fields are described in this chapter; these include electron–nuclei correlated motions, phase-adiabatic states, interwell electronic transfer, and cycle-averaged potential energy surface. These concepts play an important role in the analysis of chemical reactions under intense laser field conditions and clarify their reaction mechanisms. Simple molecules, hydrogen molecule ions, hydrogen molecules, and carbon dioxides are treated for theoretical explanations of these concepts. Then, experimental results of intense-field chemical reactions in typical molecules, benzene, fullerene, and hydrocarbons are presented.

References

1. (a) I. Kawata, H. Kono and Y. Fujimura, *Chem. Phys. Lett.* **289**, 546 (1998); (b) I. Kawata, H. Kono and Y. Fujimura, *J. Chem. Phys.* **110**, 11152 (1999).
2. (a) T. Zuo and A. D. Bandrauk, *Phys. Rev. A* **52**, R2511 (1995); (b) A. D. Bandrauk (Ed.), *Molecules in Laser Fields* (Marcel Dekker, New York, 1994).
3. R. Mulliken, *J. Chem. Phys.* **7**, 20 (1939).
4. (a) Y. Kayanuma, *Phys. Rev. B* **47**, 9940 (1993); *Phys. Rev. A* **50**, 843 (1994); (b) M. Thachuk, M. Yu. Ivanov and D. M. Wardlaw, *J. Chem. Phys.* **105**, 4094 (1996); **109**, 5747 (1998); (c) Y. Teranishi and H. Nahamura, *J. Chem. Phys.* **107**, 1904 (1997); (d) C. Zhu and H. Nakamura, *J. Chem. Phys.* **107**, 7839 (1997).
5. (a) C. Zener, *Proc. R. Soc. London. Ser. A* **137**, 696 (1932); (b) L. Landau and E. Lifshitz, *Quantum Mechanics* (Fizmatgiz, 1963).
6. A. Becker and F. H. M. Faisal, *Phys. Rev. Lett.* **84**, 3546 (2000).
7. K. Haruyama, H. Kono, Y. Fujimura, I. Kawata and A. D. Bandrauk, *Phys. Rev. A* **66**, 043403 (2002).
8. (a) S. Saugout, C. Cornaggia, A. Suzor-Weiner and E. Charron, *Phys. Rev. Lett.* **98**, 253003 (2007); (b) S. Saugout, E. Charron and C. Cornaggia, *Phys. Rev. A* **77**, 023404 (2008).
9. J. P. Nibarber, S. V. Menon and G. N. Gibson, *Phys. Rev. A* **63**, 053406 (2001).
10. C. Cornaggia, *Phys. Rev. A* **54**, R2555 (1993).
11. A. Hishikawa, A. Iwamae and K. Yananouchi, *Phys. Rev. Lett.* **83**, 1127 (1999).
12. H. Kono, S. Koseki, M. Shiota and Y. Fujimura, *J. Phys. Chem. A* **105**, 5627 (2001).
13. Y. Sato, H. Kono, S. Koseki and Y. Fujimura, *J. Am. Chem. Soc.* **125**, 8019 (2003).

14. (a) C. Cornaggia, D. Normand and J. Morellec, *J. Phys. B At. Mol. Opt. Phys.* **25**, L415 (1992); (b) C. Cornaggia, M. Schmidt and D. Normand, *Phys. Rev. A* **51**, 1431 (1995).
15. L. J. Frasinski, K. Codling and P. A. Hatherly, *Science* **246**, 1029 (1989).
16. P. Tzallas, C. Kosmidis, P. Graham, K. W. D. Ledingham, T. McCanny, S. M. Hankin, R. P. Singhal, P. F. Taday and A. J. Langley, *Chem. Phys. Lett.* **332**, 236 (2000).
17. (a) M. J. DeWitt, D. W. Peters and R. J. Levis, *Chem. Phys.* **218**, 211 (1997); (b) D. J. Smith, K. W. D. Ledingham, R. P. Singhal, H. S. Kilic, T. McCanny, A. J. Langley, P. F. Taday and C. Kosmidis, *Rapid Commun. Mass Spectrom.* **12**, 813 (1998); (c) M. Castillejo, S. Couris, E. Koudoumas and M. Martín, *Chem. Phys. Lett.* **289**, 303 (1998); (d) W. D. Ledingham, D. J. Smith, R. P. Singhal, T. McCanny, P. Graham, H. S. Kilic, W. X. Peng, A. J. Langley, P. F. Taday and C. Kosmidis, *J. Phys. Chem. A* **103**, 2952 (1999); (e) A. Talebpour, A. D. Bandrauk, K. Vijayalakshmi and S. L. Chin, *J. Phys. B At. Mol. Opt. Phys.* **33**, 4615 (2000).
18. S. Shimizu, J. Kou, S. Kawato, K. Shimizu, S. Sakabe and N. Nakashima, *Chem. Phys. Lett.* **317**, 609 (2000).
19. R. Itakura, J. Watanabe, A. Hishikawa and K. Yanamouchi, *J. Chem. Phys.* **114**, 5598 (2001).
20. (a) H. Hohmann, C. Callegari, S. Furrer, D. Grosenick, E. E. B. Campbell and I. V. Hertel, *Phys. Rev. Lett.* **73**, 1919 (1994); (b) H. Hohmann, R. Ehlich, S. Furrer, O. Kittelmann, J. Ringling and E. E. B. Campbell, *Z. Phys. D* **33**, 143 (1995); (c) R. C. Constantiescu, S. Hunsche, H. B. van Linden van den Heuvell, H. G. Muller, C. LeBlanc and F. Salin, *Phys. Rev. A* **58**, 4637 (1998); (d) N. Hay, E. Springate, M. B. Mason, J. W. G. Tisch, M. Castillejo and J. P. Marangos, *J. Phys. At. Mol. Opt. Phys.* **32**, L17 (1999); (e) J. Kou, V. Zhakhovskii, S. Sakabe, K. Nishihara, S. Shimizu, S. Kawato, M. Hashida, K. Shimizu, S. Bulanov, Y. Izawa, Y. Kato and N. Nakashima, *J. Chem. Phys.* **112**, 5012 (2000); (f) M. Tchaplyguine, K. Hoffmann, O. Dühr, H. Hohmann, G. Korn, H. Rottke, M. Wittmann, I. V. Hertel and E. E. B. Campbell, *J. Chem. Phys.* **112**, 2781 (2000); (g) I. V. Hertel, T. Laarmann, C. P. Schulz, *Adv. At. Mol. Opt. Phys.* **50**, 219 (2005); (h) T. Laarmann, C. P. Sculz and I.V. Hertel, in K. Yamanouchi, S. L. Chin, P. Agostini and G. Ferrante (Ed.), *Progress in Ultrafast Intense Laser Science*, Vol. 3 (Springer-Verlag, Berlin, 2008), p. 129.
21. (a) E. E. B. Campbell, K. Hansen, K. Hoffmann, G. Korn, M. Tchaplyguine, M. Wittmann and I. V. Hertel, *Phys. Rev. Lett.* **84**, 2128 (2000); (b) E. E. B. Campbell and R. D. Levine, *Annu. Rev. Phys. Chem.* **51**, 65 (2000); (c) F. Rohmund, M. Hedén, A. V. Bulgakov and E. E. B. Campbell, *J. Chem. Phys.* **115**, 3068 (2001).
22. V. R. Bhardwaj, P. B. Corkum and D. M. Rayner, *Phys. Rev. Lett.* **91**, 203004 (2003).
23. (a) S. Díaz-Tendero, M. Alcamí and F. Martín, *Phys. Rev. Lett.* **95**, 013401 (2005); (b) S. Díaz-Tendero, M. Alcamí and F. Martín, *J. Chem. Phys.* **123**, 184306 (2005).
24. K. Hansen, K. Hoffmann and E. E. B. Campbell, *J. Chem. Phys.* **119**, 2513 (2003).
25. A. N. Markevitch, D. A. Romanov, S. M. Smith, H. B. Schlegel, M. Y. Ivanov and R. J. Levis, *Phys. Rev. A* **69**, 013401 (2004).
26. I. Shchatsinin, T. Laarmann, G. Stibenz, G. Steinmeyer, A. Stalmashonak, N. Zhavoronkov, C. P. Schulz and I. V. Hertel, *J. Chem. Phys.* **125**, 194320 (2006).
27. (a) R. Sahnoun, K. Nakai, Y. Sato, H. Kono, Y. Fujimura and M. Tanaka, *J. Chem. Phys.* **125**, 184306 (2006); (b) K. Nakai, H. Kono, Y. Sato, N. Niitsu, R. Sahnoun, M. Tanaka and Y. Fujimura, *Chem. Phys.* **338**, 127 (2007).

28. (a) M. S. Dresselhaus, G. Dresselhaus and R. Saito, in D. M. Ginsberg (Ed.), *Physical Properties of High Temperature Superconductors IV* (World Scientific, Singapore, 1994), p. 471; (b) V. Buntar, in K. M. Kadish and R. S. Ruoff (Eds.), *Fullerenes: Chemistry, Physics, and Technology* (Wiley-Interscience, 2000), p. 691.
29. (a) A. Ballard, K. Bonin and J. Louderback, *J. Chem. Phys.* **113**, 5732 (2000); (b) K. Harigaya and S. Abe, *Phys. Rev. B* **49**, 16746 (1994).
30. R. A. Ganeev, L. B. Elouga Bom, J. Abdul-Hadi, M. C. H. Wong, J. P. Brichta, V. R. Bhardwaj and T. Ozaki, *Phys. Rev. Lett.* **102**, 013903 (2009).
31. Y. Furukawa, K. Hoshina, K. Yamanouchi and H. Nakano, *Chem. Phys. Lett.* **414**, 117 (2005).
32. K. Hoshina, Y. Furukawa, T. Okino and K. Yamanouchi, *J. Chem. Phys.* **129**, 104302 (2008).

Chapter 6

Electron Rotation Induced by Laser Pulses

In this chapter, the generation and control of an electronic angular momentum in molecules by ultrashort laser pulses are presented from the theoretical viewpoint. Results of theoretical studies on transient ring currents of π-electrons and the current-induced magnetic moments of magnesium-porphyrin and benzene with degenerate electronic excited states, which are created by circularly polarized laser pulses, are introduced. Their electronic angular momentum is transferred from photon angular momentum. A possible generation of electronic angular momentum in chiral aromatic molecules by linearly polarized laser pulses and its control are presented. Nonadiabatic effects of π-electron rotations are clarified by carrying out wavepacket simulations of chiral aromatic molecules as well. The initial rotational direction of π-electron rotations, which depend on the photon polarization direction of the laser pulse applied, affects the amplitude of subsequent molecular vibration through nonadiabatic couplings.

1. Introduction

Creation of electronic angular momentum is the fundamental process for ring current and magnetic moment.[1] Generation and control of electronic angular momentum in molecules by light is one of the fascinating research subjects since the results are applicable to molecular design of ultrafast switching devices.[2]

A circularly polarized laser field has angular momentum, *i.e.*, photon angular momentum of ± 1. It is possible to transfer the photon angular momentum to rotational angular momentum of a π-electron in a degenerate electronic excited state in a molecular system by photo-excitation. This means that transient electronic ring current in cyclic aromatic molecules can be generated by a circularly polarized laser pulse.

In this chapter, quantum mechanical simulations of electronic ring currents in Mg-porphyrin[3] and benzene[4] driven by circularly polarized femtosecond laser pulses, which are typical cyclic aromatic molecules, are presented. It is shown that the magnitude of the laser-induced ring current is larger than that obtained by the traditional magnetic induction by orders of magnitude.

We also pay attention to a new scenario for control of electronic ring current in chiral aromatic molecules by nonhelical, (i.e., linearly) polarized laser pulses.[5] It can be shown that the initial rotational direction of π-electrons in chiral molecules is determined by the polarized direction of a linearly polarized laser pulse; if clockwise rotation is the initial rotational direction of an enantiomer of a chiral molecule of interest, then counterclockwise rotation is that of the other enantiomer. That is, the initial rotational direction depends on molecular chirality. The rotational direction switches between clockwise and counterclockwise with a period. It can be also shown that unidirectional rotation, *i.e.*, clockwise or counterclockwise rotation of π-electrons is realized by applying a pair of pump and dump pulses. Effects of π-electron rotations on molecular vibration are clarified by performing wavepacket simulations of nonadiabatic dynamics of chiral aromatic molecules as well. It is found that the initial rotational direction of π-electron rotations controlled by the photon polarization direction of the laser pulse affects the amplitude of subsequent molecular vibration through nonadiabatic couplings.

2. Electronic Ring Currents Generated by Circularly Polarized Laser Pulses

2.1 *Magnesium-porphyrin*

A quantum model simulation of generation of electronic ring current in Mg-porphyrin driven by a few cycle circularly polarized laser pulse has been performed by Barth *et al.*[3] Magnesium-porphyrin possesses D_{4h} symmetry of the point group. In Fig. 1, Mg-porphyrin preoriented in the laboratory-fixed x/y-plane and an eight-cycle right circularly polarized UV π pulse propagating along the z-direction is schematically shown. A π pulse is used for population inversion in a two-level system. The system is initially in the electronic ground state. A pulsed laser with its central frequency resonant to a particular doubly degenerate state with E-symmetry is used. The doubly degenerate excited states $|E_\pm\rangle$ are expressed in terms of a linear combination of two real ones $|E_x\rangle$ and $|E_y\rangle$ as $|E_\pm\rangle = (|E_x\rangle \pm i|E_y\rangle)/\sqrt{2}$. If the system is created in $|E_+\rangle$ or $|E_-\rangle$ at time t, the time-dependent electronic current density $j(r, t)$ at position r is generated.

Figure 2 shows the results of a simulation of the electronic ring current flow induced by the right circularly polarized UV π pulse. Here, the time-dependent Schrödinger equation was solved within the three-state model with $|1^1A_{1g}\rangle$, $|4^1E_{1u+}\rangle$, and $|4^1E_{1u-}\rangle$. The corresponding molecular parameters were adapted from quantum chemical CASPT2(14/16) and TD-DFT calculations.[6]

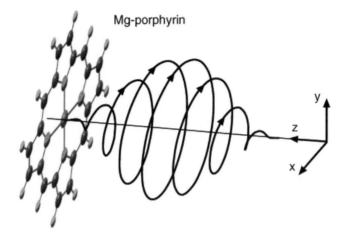

Fig. 1. Eight-cycle right circularly polarized laser pulse propagating along the z-direction to induce the unidirectional right ring current in Mg-porphyrin, preoriented in the x/y-plane. The arrows indicate the time evolution of the electric field acting on the molecule, which appears clockwise ("right") when viewed along the z-polarization direction. (Reprinted with permission from Barth et al.[3(a)] Copyright (2006) by American Chemical Society).

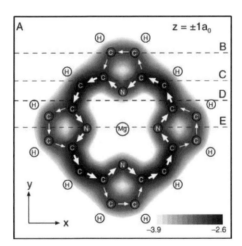

Fig. 2. The flux map of the electronic ring current in oriented Mg-porphyrin for the plane parallel at a distance $z = \pm 1a_0$. (Reproduction with permission from Barth et al.[3(a)] Copyright (2006) by American Chemical Society).

The pulse whose envelope function is $\sin^2(\pi t/t_p)$ with time duration t_p, the maximum intensity of 1.28×10^{12} W cm^{-2} and $\hbar\omega = 3.42$ eV was used.

The time-dependent electronic current density $j(r, t)$ is given as

$$j(r, t) = \frac{i\hbar}{2m_e} N \int \cdots \int [\Psi(t)\nabla\Psi^*(t) - \Psi^*(t)\nabla\Psi(t)]dr_2 \cdots dr_N d\sigma_1 \cdots d\sigma_N, \quad (1)$$

where the integration is carried out over the $N-1$ space and all the spin variables of N electrons.

The dominant contribution to the current density for $t \geq t_p$ is due to the target state, $|\Phi(t)\rangle = |4E_{u+}\rangle \exp(-iE_{4E_{u+}}t/\hbar + i\eta)$ with phase η, and the current density is expressed in a good approximation as

$$j(r) \approx j_{E+}(r) \approx j_{4e_{g+}} = \frac{i\hbar}{2m_e}[(4e_{g+})\nabla(4e_{g+})^* - (4e_{g+})^*\nabla(4e_{g+})], \quad (2)$$

where $4e_{g+}$ denotes the singly occupied excited orbital of the degenerate electronic state, $|4E_{u+}\rangle$.

The net ring current consists of two equivalent currents centered in planes parallel to the x/y-plane at $z \approx \pm 1a_0$ and vanishes in the x/y-plane. The main fluxes branch into strong and weak ones flowing along the inner C–N–C and C–C–C–C chains of bonds of the four equivalent pyrrole fragments, respectively. The electronic current of $J = 0.528\,\text{fs}^{-1}$ is obtained by integrating $j(r)$ over a half-plane, which is equivalent to $I = eJ = 84.5\,\mu\text{A}$ for circulation of π-electron around the principle molecular axis. The corresponding effective current radius, $r = 6.32\,a_0$, is obtained by taking into account the dominant contributions for the domains of the bridges between neighboring pyrrole fragments ($r_{\text{Mg-C}} = 6.45\,a_0$) plus significant ones for the inner ring ($r_{\text{Mg-N}} = 3.88\,a_0$) plus smaller ones for the outer one ($r_{\text{Mg-C}} = 8.10\,a_0$). This electronic current generates a magnetic field with maximum value of $B_z = 0.159\,\text{T}$ at the Mg atom.[3] It is interesting to compare the net electronic currents between the present values and those obtained by the diamagnetic ring current in Mg-porphyrin.[7] The value of the present net electronic currents is larger than that obtained by the traditional magnetic induction by orders of magnitude, while the effective radii of these ring currents are roughly equal each other. This shows the effectiveness of the laser-induced electronic ring currents in ultrashort transient time regimes.

2.2 Benzene

Evaluation of the photo-induced electric current of benzene driven by circularly polarized laser pulses within the time-dependent density functional theory (TDDFT) was reported by Nobusada and Yabana.[4] Benzene with D_{6h} symmetry was placed on x/y plane. In TDDFT, the time-dependent Kohn–Sham equation is solved to obtain auxiliary electronic wave functions $\psi_l(r, t)$[8]:

$$i\hbar\frac{\partial}{\partial t}\psi_l(r, t) = \left[-\frac{\hbar^2}{2m_e}\nabla^2 + V_{\text{eff}}[\rho](r, t)\right]\psi_l(r, t), \quad (3)$$

where effective potential $V_{\text{eff}}[\rho](r, t)$ is given as

$$V_{\text{eff}}[\rho](r, t) = V(r) + \int \frac{\rho(r', t)}{|r' - r|}dr' + V_{\text{xc}}[\rho](r, t) + V_{\text{int}}(r, t). \quad (4)$$

In Eq. (4), the first, second, third, and fourth terms in the left-hand side denote electron–nuclei interaction potential, time-dependent Hartree potential, exchange–correlation potential, and radiation–molecule interaction potential, respectively. In Eqs. (3) and (4), $\rho(r, t)$ is the electron density at t, which is given as $\rho(r, t) = 2\sum_{l}^{\frac{n}{2}} |\psi_l(r, t)|^2$. The adiabatic local density approximation was adopted for the exchange–correlation potential. In Eq. (4), the interaction potential between electrons and the circularly polarized electric field of laser pulse was expressed in the dipole approximation as

$$V_{\text{int}}(r, t) = exE \sin^2\left(\frac{\pi t}{T}\right) \cos(\omega t) \pm eyE \sin^2\left(\frac{\pi t}{T}\right) \sin(\omega t), \qquad (5)$$

where E is the laser field strength, T is the laser pulse duration, and ω is the central frequency of laser pulse.

Cumulative number of the electrons passing through a volume at time τ, $\sigma(\tau)$, was adopted as a measure of a ring current.[4] The cumulative number is defined as $\sigma(\tau) = \int_0^{\tau} A(t)dt$, in which $A(t)$ is the time-dependent electric current given as

$$A(t) = 2\sum_{j=1}^{\frac{N}{2}} \frac{\hbar}{2m_e i} \int_S [\psi_j^*(r, t)\nabla\psi_j(r, t) - \psi_j(r, t)\nabla\psi_j^*(r, t)]dS, \qquad (6)$$

where S denotes a surface surrounding a volume that was set to be a semicircular region ($x > 0$, $y > 0$) of the molecular ring.

Figure 3 shows the calculated cumulative number of the electrons passing through a given volume from $t = 0$ fs to 25 fs. The laser pulses with left-handed circularly polarized, resonant frequency $\hbar\omega = 6.89$ eV, which corresponds to the transition energy from $S_0(^1A_{1g})$ to $S_3(^1E_{1u})$ and pulse duration $T = 6$ fs were used. The TDDFT calculation was carried out within the frozen-chemical bond approximation, fixing the C–C bond length at 1.40 Å and the C–H bond length at 1.08 Å. The TDDFT calculation was carried out within the frozen-chemical bond approximation, fixing the C–C bond length at 1.40 Å and the C–H bond length at 1.08 Å. In Fig. 3, the cumulative numbers of the electrons in benzene are shown as a function of time for four cases with the laser intensity, $I = 1.0 \times 10^6$, 10^8, 10^9, and 10^{10} W cm^{-2}. To clarify the effects of the laser intensity, the values of the cumulative number of electrons for $I = 1.0 \times 10^6$, 10^9, and 10^{10} W cm^{-2} were multiplied by 10^{-2}, 10^{-1}, and 10, respectively. It can be seen that curves of the reduced cumulative numbers of electrons behave as that for $I = 1.0 \times 10^8$ W cm^{-2}. This means that the cumulative number is proportional to the population of the degenerate excited state $S_3(^1E_{1u})$ of benzene, in which the electronic currents are created. The time-averaged electric current is 7.0 nA for $I = 1.0 \times 10^8$ W cm^{-2}.[4] The value 7.0 nA multiplied by 1.28×10^4 is 89.6 μA. This value is close to 84.5 μA for Mg-porphyrin system driven by laser pulses with intensity of 1.28×10^{12} W cm^{-2}, which is described in

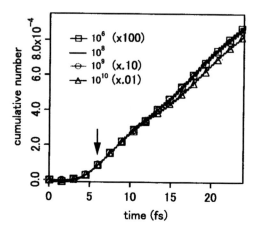

Fig. 3. Cumulative numbers of the electrons passing through a given volume of benzene from $t = 0$ to 25 fs for laser intensity of $I = 1.0 \times 10^6$, 10^8, 10^9, and 10^{10} W cm^{-2}. The central frequency of the laser is set to be a resonant transition from $S_0(^1A_{1g})$ to $S_3(^1E_{1u})$. The arrow denotes the end of the laser pulse applied. The reduced cumulative numbers for $I = 1.0 \times 10^6$, 10^9, and 10^{10} W cm^{-2} by multiplying 10^2, 10^{-1}, and 10^{-2}, respectively, are shown. Almost same behaviors of the time-dependence of the reduced cumulative number as that for $I = 10^8$ W cm^{-2} indicates the cumulative number is proportional to the intensity of the laser pulse, i.e., the population in the excited electronic state $^1E_{1u}$. (Reprinted with permission from Nobusada and Yabana.[4] Copyright (2007) by the American Physical Society).

the preceding subsection. This suggests that π-electron systems with high symmetry have the same order of magnitude in the photo-induced currents even if these dimensions are different.

3. Control of Unidirectional Rotations of π-Electrons in Chiral Aromatic Molecules

A chiral aromatic molecule has no degenerate electronic states. Therefore, any steady-state electronic ring currents cannot be created by applying a linearly polarized laser pulse without photon angular momentum. However, induction of transient rotational motion of π-electrons is possible. In this section, first it is shown that the polarization direction of a linearly polarized laser pulse determines the initial direction of π-electron rotation in a chiral aromatic molecule.[5] Then, a pump–dump control method is proposed for determining unidirectional rotation of π-electrons in a chiral aromatic molecule.[5]

A chiral aromatic molecule with a six-membered ring, 2,5-dichloro[n](3,6) pyrazinophane (DCPH; n specifies the length of the ethylene bridge $(CH_2)_n$), was chosen as shown in Fig. 4. DCPH has quasi-degenerate π-electronic excited states to create a transient angular momentum by coherent excitation. The molecule was

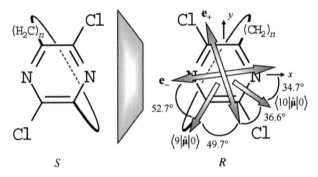

Fig. 4. S- and R-enantiomers of DCPH (2,5-dichloro[n](3,6)pyrazinophane). The directions of transition dipole moments $\langle 9|\mu|0\rangle$ and $\langle 10|\mu|0\rangle$ of an R-enantiomer are shown as well as those of photon polarization vectors $\mathbf{e}_+$, $\mathbf{e}_-$. The magnitudes of $|\langle 9|\mu|0\rangle|$ and $|\langle 10|\mu|0\rangle|$ are 2.02 ea_0 and 1.63 ea_0, respectively. (Reprinted with permission from Kanno et al.[5] Copyright (2006) by Wiley-VCH Verlag GmbH & Co. KGaA.).

assumed to be fixed to a surface by the ethylene bridge. The semi-empirical Pariser–Parr–Pople (PPP) model was used for evaluation of the electronic states.[9] In the PPP model, DCPH is regarded as a system having eight p_z orbitals and 10 π-electrons. The aromatic ring forms a planar structure, and the C–C and C–N bond lengths are set to be 1.40 Å, the C–Cl bond length 1.80 Å, and a(N–C–Cl) = 120°.

The time-dependent Schrödinger equation was solved by expanding electronic wavepacket $|\Psi(t)\rangle$ in terms of 136 singlet eigenstates of the π-electronic Hamiltonian obtained at the level of singly and doubly excited configuration interaction (SDCI).[5] In this model, π-electron rotation can be quantified by the momentum expectation value $p(t) \equiv \langle\Psi(t)|\hat{p}|\Psi(t)\rangle$, where $\hat{p}$ is defined using the momentum of complex molecular orbitals for a circular path along an aromatic ring. Here, in a circular motion of a particle, angular velocity of the particle is equivalent to its velocity divided by the radius of the circle. Thus, the rotational angle of π-electrons $\theta(t)$ is defined as

$$\theta(t) = \frac{1}{m_e b}\int_0^t d\tau p(\tau), \quad (7)$$

where b is the radius of the ring. The expectation values $p(t)$ and $\theta(t)$ are used as measures of π-electron rotation.

Let us consider quasi-degenerate π-electronic excited states $|9\rangle = |5^1 B_u\rangle$ and $|10\rangle = |6^1 B_u\rangle$ with the energy gap $2\Delta E \equiv E_{10} - E_9 = 0.105$ eV ($E_9 - E_0 = 7.66$ eV). The transition dipole moments $\langle 9|\mu|0\rangle$ and $\langle 10|\mu|0\rangle$ of an R-enantiomer of DCPH are illustrated in Fig. 4 and those of an S-enantiomer are their mirror images. Linear combinations of $|9\rangle$ and $|10\rangle$ give the approximate eigenstates of the angular momentum operator $l = b\hat{p}$, denoted as $|+\rangle$ and $|-\rangle$, where $\langle\pm|l|\pm\rangle = \pm 0.862\hbar$. For an R-enantiomer, this is described as

$$|\pm\rangle = 2^{-\frac{1}{2}}[|9\rangle \pm i|10\rangle], \quad (8)$$

For an S-enantiomer, the sign $\pm$ on the right-hand side of Eq. (8) is replaced by $\mp$. π-Electrons with positive (negative) angular momentum travel counterclockwise (clockwise) around the ring in Fig. 4.

Let us now design a linearly polarized laser pulse to generate predominantly either $|+\rangle$ or $|-\rangle$. First, the central frequency ω of the pulse is set as $\omega = (E_9 + \Delta E - E_0)/\hbar$. Next, the polarization direction of $E(t)$ is determined by its alignment relative to $\langle 9|\boldsymbol{\mu}|0\rangle$ and $\langle 10|\boldsymbol{\mu}|0\rangle$. The polarization unit vector of the pulse is chosen in two ways, $\mathbf{e}_+$ or $\mathbf{e}_-$ defined as $\langle 9|\boldsymbol{\mu}|0\rangle\mathbf{e}_\pm = \mp\langle 9|\boldsymbol{\mu}|0\rangle\mathbf{e}_\pm$ for each enantiomer. The directions of $\mathbf{e}_\pm$ for an R-enantiomer are illustrated in Fig. 4. At the moment of irradiation ($t = t_i$), the pulse with $\mathbf{e}_+$ ($\mathbf{e}_-$) produces an in-phase superposition $|9\rangle + |10\rangle$ (out-of-phase superposition $|9\rangle - |10\rangle$) in $|\Psi(t_i)\rangle$. At $t > t_i$, the electron wavepacket propagates freely. Hence, $|9\rangle \pm |10\rangle$ in $|\Psi(t_i)\rangle$ temporally evolves as $|9\rangle \pm \exp[-i\phi(t)]|10\rangle$ except for the global phase factor, where $\phi(t) = 2\Delta E(t - t_i)/\hbar$. $\exp[-i\phi(t)]$ changes as $+1 \to -i \to -1 \to +i \to +1 \to \cdots$ with the progression of $t - t_i, 0 \to T/4 \to T/2 \to 3T/4 \to T \to \cdots$, where $T \equiv \pi\hbar/\Delta E$. This indicates that, in the first quarter period of T after excitation, $|\mp\rangle$ is created in an R-enantiomer, while $|\pm\rangle$ is created in an S-enantiomer. Thus, the polarization direction determines the initial direction of π-electron rotation; the rotation direction switches between clockwise and counterclockwise with the period T.

If a molecule is highly symmetric, e.g., benzene, $\exp[-i\phi(t)]|10\rangle$ takes an infinite time to reach $-i$ since $\Delta E = 0$ for the optically active doubly degenerate excited states of benzene. That is, lowering the molecular symmetry is essential for the selective generation of either $|+\rangle$ or $|-\rangle$ by a linearly polarized laser pulse.

Figure 5 shows results of a numerical simulation of the electronic dynamics for an R-enantiomer driven by a linearly polarized π pulse that is designed to initially create $|-\rangle$. The polarization vector of the pulse is $\mathbf{e}_+$. Figure 5(a) shows the electric field of the applied pulse with form $\boldsymbol{E}(t) = \mathbf{e}_+ E_0 \sin^2(\pi t/t_d) \cos(\omega t)$ for $0 < t < t_d$ and otherwise $\boldsymbol{E}(t) = 0$. Here, the laser parameters were $\hbar\omega = 7.72$ eV, $E_0 = 1.63$ GVm^{-1}, and $t_d = 26.6$ fs. In Fig. 5(b), the solid, dotted, and dashed-dotted lines denote the temporal behaviors in the populations of $|0\rangle$, $|+\rangle$, and $|-\rangle$, respectively. The values for $p(t)$ and $\theta(t)$ are plotted in Figs. 5(c) and 5(d), respectively. If the pulse duration t_d is less than the oscillation period T, $|9\rangle + |10\rangle$ is mainly produced around the pulse peak, i.e., $t_i = t_d/2 = 13.3$ fs. At $t > 13.3$ fs, a significant amount of the population is transferred to $|9\rangle - i|10\rangle$, i.e., $|-\rangle$ in Fig. 5(b), and accordingly π-electrons start to rotate clockwise (see Figs. 4 and 5(c)). The total population of π-electrons in $|+\rangle$ and $|-\rangle$ reaches 0.908 when the laser pulse ceases at $t = 26.6$ fs. From the energy–time uncertainty relation, the final value of the total population is maximum at the pulse duration $t_d = 26.6$ fs. A smaller bandwidth of a longer pulse does not cover the energy gap $2\Delta E$ sufficiently, and, on the other hand, a broader bandwidth of a shorter pulse populates other excited states. At $t > 26.6$ fs,

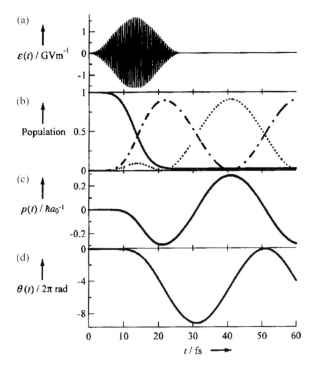

Fig. 5. The initial direction of π-electron rotation in chiral aromatic molecule DCPH, which is generated by a one-color linearly polarized π laser pulse. (a) Linearly polarized laser pulse to initially create $|-\rangle$ of an R-enantiomer. (b) Temporal behavior in the populations of $|0\rangle$ (solid line), $|+\rangle$ (dotted line), and $|-\rangle$ (dashed-dotted line). (c) Expectation value of momentum $p(t)$. (d) Expectation value of rotational angle $\theta(t)$. (Reprinted with permission from Kanno et al.[5] Copyright (2006) by Wiley-VCH Verlag GmbH & Co. KGaA.).

the population of 0.908 is exchanged between $|+\rangle$ and $|-\rangle$ as expected. $p(t)$ and $\theta(t)$ thus oscillate with the period of $T = 39.5$ fs, and π-electrons are estimated to circulate around the ring more than nine times within this period (see Fig. 5(d)).

It should be noted that the direction of the created angular momentum vector of electrons driven by circularly polarized laser field is defined in the laboratory flame, while the direction of the angular momentum created by linearly polarized one is defined by the molecular flame.

Let us consider an intuitive and pump–dump control method for determining unidirectional rotation of π-electrons.[5] It should be noted that the population created in $|-\rangle$ can be dumped to $|0\rangle$ by applying a dump pulse with $\mathbf{e}_-$ just after the population has completely shifted to $|9\rangle - |10\rangle$. Figure 6 shows the results of a pump–dump control simulation of an R-enantiomer. Figure 6(a) shows the pulse forms: for the pump pulse, $\hbar\omega = 7.72$ eV, $E_0 = 2.24$ GVm^{-1}, and $t_d = 19.4$ fs were adopted; for the dump pulse, $\hbar\omega = 7.72$ eV, $E_0 = 2.37$ GVm^{-1}, and $t_d = 19.4$ fs were adopted. The polarization vectors of the pump and dump pulses are $\mathbf{e}_+$ and $\mathbf{e}_-$, respectively.

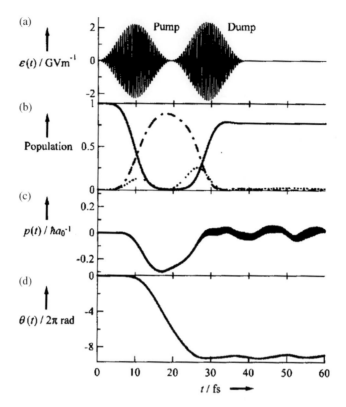

Fig. 6. Generation of unidirectional π-electron rotations in an R-enantiomer of DCPH by a pump–dump pulse method. (a) Pump and dump pulses for clockwise rotation. (b) Temporal behavior in the populations of $|0\rangle$ (solid line), $|+\rangle$ (dotted line), and $|-\rangle$ (dashed-dotted line). (c) Expectation value of momentum $p(t)$. (d) Expectation value of rotational angle $\theta(t)$. (Reprinted with permission from Kanno et al.[5] Copyright (2006) by Wiley-VCH Verlag GmbH & Co. KGaA.).

Around the peak of the dump pulse at $t = 29.1$ fs, the populations of $|+\rangle$ and $|-\rangle$ are nearly equal; in other words, an out-of-phase superposition $|9\rangle - |10\rangle$ is created (see Fig. 6(b)). At $t > 29.1$ fs, most of the population is dumped to $|0\rangle$; the rest is brought to higher excited states. Consequently, the value of $p(t)$ is almost zero, and reverse rotation is successfully prevented as shown in Figs. 6(c) and 6(d)). Thus, a pair of pump and dump pulses realizes unidirectional rotation of π-electrons. Moreover, repetition of the unidirectional rotation can be achieved by a sequence of pulse pairs.

π-Electron rotation in an S-enantiomer can be controlled in the same way. By reversing the polarization directions of the pump and dump pulses for an R-enantiomer with respect to a reflection plane, π-electrons in an S-enantiomer are rotated counterclockwise shown in Fig. 6.

So far control scenario of π-electron rotations in chiral aromatic molecules was based on a pump–dump pulse method. Nonintuitive and systematic scenario that is based on an optimal control theory can be adopted to obtain as much angular momentum as possible for a clockwise or counterclockwise circular path along an aromatic ring at the end of control by linearly polarized laser fields.[10] Optimal control of electron rotations in two-dimensional (2D) systems in the fields of semiconductor quantum rings has been also proposed for the construction of a coherent laser-driven single-gate qubit.[11]

4. Nonadiabatic Effects of Laser-Induced π-Electron Rotation

In Secs. 2 and 3, photo-induced electron dynamics was theoretically treated under the nuclei-fixed condition. The electronic and nuclear motions are coupled to each other when π-electron rotation lasts as long as the period of molecular vibrations (several tens of femtoseconds). It is well known that quasi-degenerate states are strongly coupled each other by molecular vibrations.[12] It is also well recognized that nonadiabatic couplings play an essential role in polyatomic molecules even in isolated conditions.[13] Therefore, π-electron rotation should be subjected to undergo nonnegligible nonadiabatic perturbations by nuclear motions.

In this section, the effects of π-electron rotation on molecular vibration are clarified by carrying out wavepacket simulations of nonadiabatic dynamics in a chiral aromatic molecule irradiated by a linearly polarized laser pulse.[14] The simplified model molecule, 2,5-dichloropyrazine (called DCP), in which the ansa group, the ethylene bridge $(CH_2)_n$, of chiral 2,5-dichloro[n](3,6)pyrazinophane (DCPH), is replaced with hydrogen atoms was used. This simplification is valid because π-electrons cannot be directly affected by the ansa group with σ-electrons, and the directions of the transition moments are unchanged.

DCP is of C_{2h} symmetry at both the optimized geometry of the ground state $|G\rangle \equiv |1^1A_g\rangle$ and those of a pair of optically allowed quasi-degenerate excited states, $|L\rangle \equiv |3^1B_u\rangle$ and $|H\rangle \equiv |4^1B_u\rangle$. Here, $|L\rangle(|H\rangle)$ corresponds to $|9\rangle(|10\rangle)$ of DCPH. Vibrational modes with displacements from the optimized geometry of $|G\rangle$ to that of $|L\rangle$ and $|H\rangle$ are the totally symmetric modes. Furthermore, the vibrational modes that couple two ^1B_u states are also totally symmetric A_g modes. A_g breathing and distortion modes (Fig. 7) with large potential displacements and nonadiabatic coupling matrix elements were taken into account as the coupling modes. The ground-state harmonic wave numbers of the breathing and distortion modes are 1160 and 1570 cm^{-1}, respectively.

Nonadiabatic coupling effects on the wavepacket propagation were quantum-mechanically treated. The state vector of the system was expanded in terms of the three diabatic states $\{|M^D\rangle\}$, constructed as a linear combination of the three adiabatic states $|G\rangle$, $|L\rangle$, and $|H\rangle$. The time evolution of the expansion coefficients

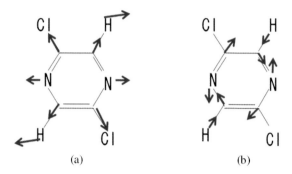

Fig. 7. Effective coupling modes of DCP (2,5-dichloropyrazine): (a) breathing mode and (b) distortion mode.

for $|M^D\rangle$, $\psi_M^D(\mathbf{Q}, t)$, where $\mathbf{Q}$ is the 2D normal coordinate vector, was obtained by numerically solving the coupled equations with $M = L$ and H:

$$i\hbar \frac{\partial}{\partial t}\psi_M^D = -\left(\frac{\hbar^2}{2}\right)\nabla^2 \psi_M^D + \sum_{M'}[V_{MM'}^D(\mathbf{Q}) - \boldsymbol{\mu}_{MM'}^D(\mathbf{Q}) \cdot \mathbf{E}(t)]\psi_{M'}^D, \quad (9)$$

where ∇^2 is the Laplacian with respect to $\mathbf{Q}$. $V_{MM'}^D(\mathbf{Q})$ involves the diabatic potentials and couplings, and $\boldsymbol{\mu}_{MM'}^D(\mathbf{Q})$ is the transition moment between the two diabatic states. The resultant diabatic wavepacket $\psi_M^D(\mathbf{Q}, t)$ is converted to adiabatic wavepacket $\psi_M(\mathbf{Q}, t)$.

Figures 8(a) and 8(b) show the temporal behavior in the expectation value of electronic angular momentum $L(t)$ and that of vibrational coordinate $\mathbf{Q}(t)$, respectively, by applying a laser pulse with the linear polarization vector $\mathbf{e}^+$ and that with $\mathbf{e}^-$ (hereafter termed $\mathbf{e}^+$ and $\mathbf{e}^-$ excitations). The initial nuclear wavepacket was set to be the vibrational ground-state wavefunction of $|G\rangle$. The pulse duration of a $\sin^2$ envelope was set at 7.26 fs. The peak intensities of the pulses with $\mathbf{e}^+$ and $\mathbf{e}^-$ were set to 5.53 and 9.02 GVm^{-1}, respectively.

It can be clearly seen from Fig. 8(a) that the amplitudes of $L(t)$ gradually decay for both cases. The decay of the angular momentum originates from two factors: one is the decrease of the overlap between the wavepackets moving on the relevant two adiabatic potential energy surfaces and the other is electronic relaxations due to nonadiabatic couplings. The former factor occurs even within the Born–Oppenheimer approximation. The latter is the major factor and is absent in a frozen-nuclei model.

It should be noted in Fig. 8 that there are significant differences between the oscillatory decays of the angular momentum for $\mathbf{e}^+$ and $\mathbf{e}^-$ excitations. The curve of $L(t)$ for $\mathbf{e}^+$ excitation can be approximately expressed in a sinusoidal exponential decay form with its oscillation period of ∼9.4 fs and lifetime of ∼7 fs. In contrast, the amplitude of $L(t)$ for $\mathbf{e}^-$ excitation does not undergo a monotonic decrease but

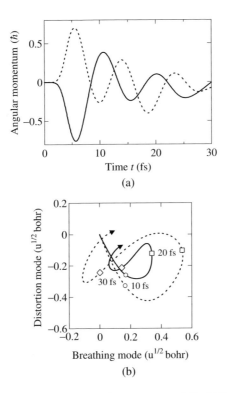

Fig. 8. (a) Expectation value of electronic angular momentum $L(t)$. (b) Expectation value of vibrational coordinate $Q(t)$. The solid and dotted lines denote the expectation values for $\mathbf{e}_+$ and $\mathbf{e}_-$ excitations, respectively. The positions of the expectation values at 10, 20, and 30 fs are indicated by circle, rectangle, and declined rectangle, respectively. The values of $Q(t)$ are plotted up to $t = 40$ fs. The laser pulse ceases at $t = 7.26$ fs. (Reprinted with permission from Kanno et al.[14] Copyright (2010) by the American Physical Society).

makes a small transient recovery around $t \sim 14$–20 fs. Its oscillation period is a little shorter than that for $\mathbf{e}^+$ excitation in this time range. The behaviors in the oscillation period of the angular momentum for $\mathbf{e}^+$ and $\mathbf{e}^-$ excitations originate from the difference in the energy gap between the two adiabatic potential energy surfaces for the regions in which the wavepackets run.

An interesting fact is that the behaviors of $Q(t)$ in Fig. 8 are strongly dependent on the polarization of the applied pulse. The amplitude of $Q(t)$ for $\mathbf{e}^-$ excitation is more than two-times larger than that for $\mathbf{e}^+$ excitation. This finding is remarkable in the sense that the initial rotation direction of π-electrons controlled by the polarization direction of the laser pulse greatly affects the amplitude of subsequent molecular vibration through nonadiabatic couplings. This indicates that molecular chirality can be identified by applying time-resolved vibrational spectroscopic methods such as transient impulsive Raman scattering since the rotation direction of π-electrons differs between enantiomers according to their alignments relative to the polarization direction.

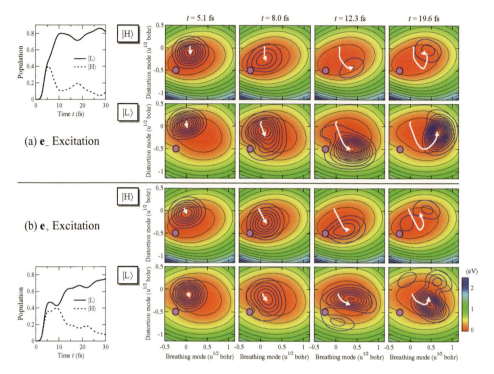

Fig. 9. Left panel: Temporal behavior in the populations of $|L\rangle$ (solid line) and $|H\rangle$ (dotted line). Right panels: Propagation of the adiabatic wavepackets on the 2D adiabatic potential energy surfaces of $|L\rangle$ and $|H\rangle$. The origin of the potential energy surfaces is taken to be the optimized geometry of $|G\rangle$. The bold contours represent the probability densities $|\psi_L(Q,t)|^2$ and $|\psi_H(Q,t)|^2$ and the arrows indicate the motion of the center of the wavepackets. The avoided crossing is signified by a circle. (Reprinted with permission from Kanno et al.[14] Copyright (2010) by The American Physical Society).

Figures 9(a) and 9(b) show temporal behaviors in the population and wavepacket dynamics on the relevant two adiabatic potential energy surfaces after e^+ and e^- excitations, respectively. For e^- excitation, the probability densities $|\psi_L(Q,t)|^2$ and $|\psi_H(Q,t)|^2$ at $t \sim 5$ fs have almost the same shape as that of the initial wavepacket, while the two excited wavepackets created are out-of-phase from the definition of e^-. As the wavepackets start to move along the gradient of each potential energy surface, significant population transfer occurs from $|H\rangle$ to $|L\rangle$ by nonadiabatic transition. Consequently, the population of $|L\rangle$ at $t \sim 10$ fs is more than seven times larger than that of $|H\rangle$, although they are almost equal at $t \sim 5$ fs. The loss of a superposition of $|L\rangle$ and $|H\rangle$ reduces the amplitude of $L(t)$ as in Fig. 8(a). The direction of the population transfer is reversed periodically despite the rather small amount of the population transferred; the revival of the superposition of $|L\rangle$ and $|H\rangle$ around $t \sim 14$–20 fs brings about the transient recovery of the angular momentum. $\psi_L(Q,t)$ moves in the high-potential energy region following the potential gradient of $|L\rangle$, which leads to the large-amplitude vibration in Fig. 8(b).

In this chapter, theoretical investigations on ultrafast laser-induced rotational dynamics of π-electrons in aromatic molecules are presented. Two types of aromatic molecules with different nature, achiral and chiral ones, are adopted as a model system and two types of laser fields with helical and nonhelical ones are adopted. It has been shown by simulations that electron currents and the magnetic moments of nonhelical molecules with aromatic ring, which are generated by helical (circularly polarized) laser pulses, become larger than those generated by the traditional magnetic induction by orders of magnitude. The initial direction of π-electron rotation in a chiral aromatic molecule can be induced by a nonhelical (linearly polarized) laser pulse. The initial direction of π-electron rotation depends on the polarization direction of a linearly polarized laser pulse. If π-electrons of S-enantiomer start to rotate clockwise by applying a laser pulse with a polarization direction properly chosen, then those of R-enantiomer start to rotate counterclockwise by applying the laser pulse with the same polarization direction. This suggests a new method to identify molecular chirality since the directions of the induced electron currents and magnetic moments are determined by the chiral molecule of interest not by applied laser pulses. It has been also shown that π-electron in a chiral aromatic molecule can be consecutively rotated by applying a sequence of linearly polarized pump and dump pulses whose directions are properly chosen. Finally, nonadiabatic couplings between π-electron rotations and molecular vibrations in a chiral aromatic molecule irradiated by a linearly polarized laser pulse are examined by carrying out wavepacket simulations. The vibrational amplitudes strongly depend on the initial rotational direction, clockwise or counterclockwise. This indicates that attosecond π-electron rotations may be observed by spectroscopic determination of femtosecond vibrations such as transient impulsive Raman spectroscopy.

References

1. (a) Y. R. Shen, *The Principles of Nonlinear Optics* (John Wiley, New York and Chichester, 1984); (b) P. Lazzeretti, *Prog. Nucl. Magn. Reson. Spectrosc.* **36**, 1 (2000); (c) E. Steiner and P. W. Fowler, *J. Phys. Chem. A* **105**, 9553 (2001).
2. (a) P. Krause, T. Klamroth and P. Saalfrank, *J. Chem. Phys.* **123**, 074105 (2005); (b) P. Král and T. Seideman, *J. Chem. Phys.* **123**, 184702 (2005); (c) J. Jortner and M. Ratner (Eds.), *Molecular Electronics* (Blackwell, Oxford, 1997); (d) A. M-Abiague and J. Berakdar, *Phys. Rev. Lett.* **94**, 166801 (2005); (e) S. Abe, J. Yu and W. P. Su, *Phys. Rev. B* **45**, 8264 (1992).
3. (a) I. Barth, J. Manz, Y. Shigeta and K. Yagi, *J. Am. Chem. Soc.* **128**, 7043 (2006); (b) I. Barth and J. Manz, *Angew. Chem. Int. Ed.* **45**, 2962 (2006).
4. K. Nobusada and K. Yabana, *Phys. Rev. A* **75**, 032518 (2007).
5. M. Kanno, H. Kono and Y. Fujimura, *Angew. Chem. Int. Ed.* **45**, 7995 (2006).
6. M. Rubio, B. O. Roos, L. S-Andrés and M. Merchán, *J. Chem. Phys.* **110**, 7202 (1999).

7. E. Steiner, A. Soncini and P. W. Fowler, *Org. Biomol. Chem.* **3**, 4053 (2005).
8. (a) E. Runge and E. K. U. Gross, *Phys. Rev. Lett.* **52**, 997 (1984); (b) R. G. Parr and W. Yang, *Density-Functional Theory of Atoms and Molecules* (Oxford University Press, Oxford, 1989).
9. M. Suzuki and S. Mukamel, *J. Chem. Phys.* **120**, 669 (2004).
10. M. Kanno, K. Hoki, H. Kono and Y. Fujimura, *J. Chem. Phys.* **127**, 204314 (2007).
11. E. Räsänen, A. Castro, J. Werschnik, A. Rubio and E. K. U. Gross, *Phys. Rev. Lett.* **98**, 157404 (2007).
12. H. Köppel, W. Domcke and L. S. Cederbaum, *Adv. Chem. Phys.* **57**, 59 (1984).
13. (a) S. H. Lin, *J. Chem. Phys.* **44**, 3759 (1966); (b) M. Bixon and J. Jortner, *J. Chem. Phys.* **48**, 715 (1968).
14. M. Kanno, H. Kono, Y. Fujimura and S. H. Lin, *Phys. Rev. Lett.* **104**, 108302 (2010).

Chapter 7

Photoisomerization and Its Control

In this chapter, two fundamental aspects of ultrafast photoisomerization are presented. The first one is concerned with observation of real-time structural evolution during *cis–trans* isomerization of stilbene in solution, which is a typical example of real-time observation of photoisomerization. Here, a nuclear wavepacket is created in an S_1 excited state by impulsive Raman processes after a pump pulse is turned off. The other aspect is concerned with optical control of photoisomerization. An experiment on coherent control of *trans–cis* photoisomerization of retinal in bacteriorhodopsin and a quantum control of retinal isomerization in rhodopsin are briefly described. The optical control experiment was carried out under a weak field condition so as to keep the biological reaction system undamaged.

1. Introduction

Photoisomerization is a general term for rearrangement of a molecule in an electronic excited state. Among photoisomerizations, *trans–cis* (*cis–trans*) photoisomerization has been investigated in detail since it plays an important role in biomolecules as well as in ultrashort switching devices.[1,2] For example, photo-induced isomerization of retinal triggers biological functions such as vision or phototaxis.[3] Isomerization of retinal takes place at a specific carbon–carbon double bond, whose reaction site depends on the protein surrounding the retinal: in visual pigment rhodopsin, 11-*cis–trans* isomerization takes place in the S_1 excited state (Fig. 1), while all *trans*- to 13-*cis*-photoisomerizations occur in bacteriorhodopsin (bR).[4,5]

There have been many experimental and computational works on the mechanism of ultrafast *trans–cis* (*cis–trans*) photoisomerization of unsaturated hydorcarbons.[6] The photoisomerization is induced by nonadiabatic processes involving conical intersection between relevant two electronic states (ground and excited electronic states).[7] Femtosecond transition-state dynamics of *cis*-stilbene under an isolated condition has already been reported by Zewail *et al.*[8] In this chapter, the real-time observation of structural evolution during stilbene photoisomerization in solution,

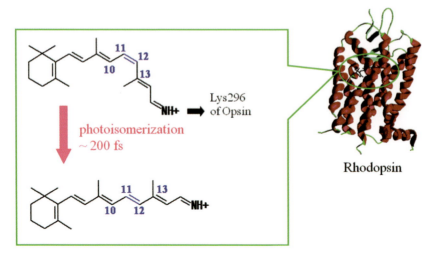

Fig. 1. An illustration of 11-*cis*- to all *trans*-photoisomerizations of retinal in the visual pigment rhodopsin. 11-*cis*-retinal chromophore is bound to a lysine residue (Lys-296) *via* a protonated Schiff base linkage.

which has been reported by Tahara's group,[9] is first presented as a typical example of real-time observation of ultrafast reactions in an electronic excited state in condensed phase. Here, the temporal behaviors of the nuclear wavepacket created in the S_1 excited state by impulsive Raman processes were measured and the behaviors were analyzed using a computational approach based on *ab initio* quantum mechanical theory. Next, experiments on optical (coherent) control of *trans–cis* photoisomerization in retinal in bacteriorhodopsin are briefly presented.[10] Here, a genetic algorithm and feedback approach were adopted for maximization or minimization of the reaction yield. The experiment was carried out under a weak field condition to avoid the production of by-products such as ionized species. Finally, quantum control theory is applied to retinal isomerization in rhodopsin in order to clarify behaviors of the wavepacket near the conical intersection, which plays an essential role in photoisomerization.[11]

2. Real-Time Observation of Stilbene *Cis–Trans* Isomerization

Real-time structural evolution during stilbene photoisomerization in solution has recently been observed by means of femtosecond transient impulsive Raman spectroscopy by Takeuchi *et al.*[9] The schematic picture of the observation system is shown in Fig. 2(a). Here, the laser system consists of a sequence of three pulses (P_1; 267 nm, 150 fs; P_2; 620 nm, 11 fs, and P_3; 620 nm, 11 fs). The first pulse (P_1), pump pulse, creates a population in an excited state S_1 of *cis*-stilbene; after a delay of ΔT,

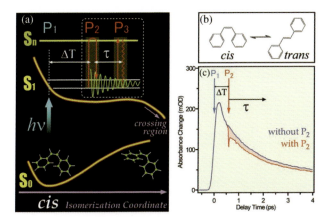

Fig. 2. (a) Schematic illustration of observation of *cis–trans* isomerization of stilbene in the S_1 state by applying a sequence of three laser pulses. After a delay ΔT, the second pulse (P_2) is applied to generate a nuclear wavepacket in the S_1 state. The third pulse (P_3) is used to monitor time-resolved $S_n \leftarrow S_1$ absorption signals. τ, delay time for impulsive Raman measurements. (b) *Cis–trans* photoisomerization of stilbene. (c) Typical time-resolved traces of the $S_n \leftarrow S_1$ absorption of *cis*-stilbene measured with and without the P_2 pulse. The difference between the two traces gives a time-resolved impulsive Raman signal. mOD, milli-optical density. (Reprinted with permission from Takeuchi *et al.*[9] Copyright (2008) by American Association for the Advancement of Science).

the second pulse P_2 resonant with a higher excited-state (S_n) transient absorption impulsively induces a vibrational coherence (nuclear wavepacket) of Raman active modes in S_1 and gives the driving force of the wavepacket motion along the reaction coordinate of *cis–trans* isomerization. The third pulse (P_3) monitors the excited-state absorption, the intensity of which is modulated by the nuclear wavepacket motion.

Time-resolved impulsive Raman signals were measured at three delay times, $\Delta T = 0.3$, 1.2, and 2 ps, as shown in Fig. 3(a). Figure 3(b) shows that the central frequency of the 240 cm^{-1} vibrational mode decreases in amount of 21 cm^{-1} as ΔT increases from 0.3 ps to 1.1 ps. This indicates that the mode changes while the isomerization proceeds. The change is also supported from the inset of Fig. 3(a) in which the intensity maxima of the three beating components gradually shift in time, which means a lengthening of the oscillation period with increasing delay. The 240 cm^{-1} mode, whose vibrational period is approximately equal to 140 fs, probes the structural evolution of *cis*-stilbene as a spectator (Fig. 4(a)).

To analyze the observation with actual structural changes in S_1 *cis*-stilbene, potential energy surfaces and instantaneous mode frequency of 240 cm^{-1} were calculated by using density functional theory (DFT) and time-dependent DFT theory.[9] The gross feature of the S_1 potential energy surface is shown in Fig. 4(b) in terms of Q (isomerization coordinate) and q (spectator coordinate), in which the minimum energy path denoted by a curved arrow starts from the Franck–Condon (FC) position

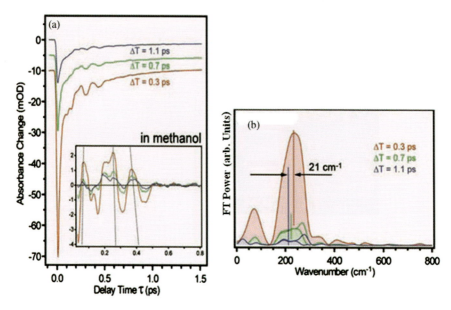

Fig. 3. (a) Time-resolved impulsive Raman signals at three delay times, $\Delta T = 0.3$, 1.2, and 2 ps, of *cis*-stilbene in methanol with a *cis*-stilbene concentration of 0.02 mol dm^{-3}. The insert shows the beating components after subtraction of the population components. The largest beating component refers to the impulsive Raman signal at $\Delta T = 0.3$. The lines connect the corresponding maxima of the three beating components. The slopes of the lines indicate that the oscillation period becomes longer with increase in ΔT. (b) Fourier transform (FT) power spectra of the beating components obtained at the three delay times. The largest signal at $\Delta T = 0.3$ ps appears around 240 cm^{-1}, which is characteristic of S$_1$ *cis*-stilbene. The vibration has been assigned to a mode that involves the motion of phenyl-C=C bending, ethylenic C=C torsion and phenyl torsion. (Reprinted with permission from Takeuchi *et al.*[9] Copyright (2008) by American Association for the Advancement of Science).

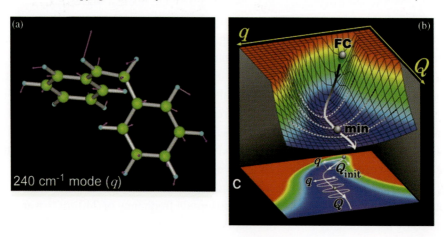

Fig. 4. (a) Nuclear motions of the 240 cm^{-1} mode at the geometry near the energy minimum on the S$_1$ potential surface. (b) Schematic illustration of the S$_1$ potential surface of *cis*-stilbene. The reaction coordinate is denoted by Q and the spectator mode of the *cis–trans* isomerization is denoted by q. (Reprinted with permission from Takeuchi *et al.*[9] Copyright (2008) by American Association for the Advancement of Science).

(Q_{init}) toward a conical intersection region between S_1 and S_0 states *via* the minimum energy point (**min**) of S_1. It should be noted that the potential curvature of the spectator (q) becomes smaller, as the molecule evolves along the isomerization coordinate. The calculated time-dependent behaviors of mode q are consistent with those in the impulsive Raman experiment (Fig. 3).

Solvent-dependence in isomerization rates has also been investigated.[9,12] In this section, photoisomerization of *cis*-stilbene in methanol is presented. Experiments on the photoisomerization in a nonpolar solvent, hexadecane, have been carried out.[9] The temporal change in the frequency of the spectator mode depends on the solvent: the rate of the frequency down-shift is higher in methanol than in hexadecane. With this change, the isomerization rate in methanol increases by a factor of 2.7 (0.77 to 2.08 ps^{-1}).

3. Quantum Control of Retinal Isomerization in Bacteriorhodopsin

Real-time spectroscopy of transition states of retinal isomerization from the all-*trans* to the 13-*cis* conformations in bacteriorhodopsin (bR) as shown below and theoretical treatments for the isomerization mechanisms have already been reported by many groups.[13] Photoisomerization of the retinal chromophore in bR occurs with a relatively high quantum yield of 65%.

This system has been adopted as a target to demonstrate an quantum (coherent) control experiment on condensed phases by Prokhorenko *et al.*[10] They used a genetic algorithm and feedback approach to optimize the isomerization yield, as shown in Chap. 1. They reported manipulation of the absolute yield of 13-*cis* retinal over a 40% range, *i.e.*, selective enhancement or suppression of the isomerization yield by 20% in either direction. The coherent experiment has been carried out in a weak field regime to ensure that the resultant dynamics would pertain to the protein behavior under normal functional conditions (where only 1 of 300 retinal molecules absorbs a photon during the excitation cycle).

Figure 5 shows the effects of evolving pulse shapes on the product of 13-*cis*-retinal in bR, which is measured by 630-nm absorption: (a) and (b) under the control condition of giving maximization and minimization of the isomerization yield, respectively. The product, termed K intermediate of 13-*cis*-retinal, has a well-resolved positive differential absorption (dA) band in the 630- to 640-nm

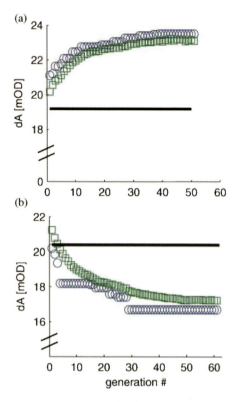

Fig. 5. Effect of evolving pulse shapes on the isomerization yield of retinal in bR, measured at 630 nm delayed 20 ps after excitation. In (a) maximization and (b) minimization experiments, squares correspond to averaged values of the differential absorbance dA (over the whole population), and circles correspond to the most effective pulses [for enhancement in (a) or suppression in (b)] within the current population. The solid horizontal lines correspond to the dA measured by excitation with the transform-limited pulse (19 fs FWHM) as a baseline for comparison. (Reprinted with permission from Prokhorenko et al.[10] Copyright (2006) by American Association for the Advancement of Science).

range. Different excitation pulse shapes were generated by appropriate manipulation of incoming transform-limited pulses with 19 fs full-width-at-half-maximum (FWHM), centered at 565 nm with band width of 60 nm in both frequency and pulse domains. The optimization algorithm was used for a starting set of 30 pulses with randomly distributed spectral phases and amplitudes. Here, pulse energies were 16–17 nJ, corresponding to a fluence of 2.7×10^{14} photons cm^{-1}.

Figure 6(a) shows the spectrum of the optimal excitation pulse for the maximization of isomerization yield; the gravity center is at ~557 nm and there is a regular structure composed of several peaks spaced ~6 nm apart.

The anti-optimization experiments (Fig. 6(d)) yielded a pulse that suppressed isomerization efficiency by 24%. The corresponding spectrum was relatively broad and red-shifted to 577 nm. Frequency-resolved optically gated (FROG) measurements

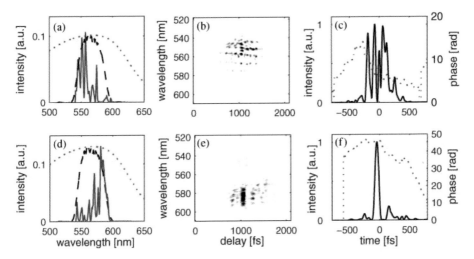

Fig. 6. (a) Spectral profile of the optimal excitation pulse (solid line), the transform-limited (19 fs FWHM) pulse (dashed line), and bR transmittance spectrum (dotted line). (b) Measured FROG trace for the optimal pulse. (c) Retrieved intensity (solid line) and phase profile (dashed line) from the FROG trace. (d–f) Corresponding data for the anti-optimal pulse. a.u., arbitrary units; rad, radians. (Reprinted with permission from Prokhorenko et al.[10] Copyright (2006) by American Association for the Advancement of Science).

were carried out to determine the temporal profiles of the optimized pulses (Figs. 6(b) and 6(e)). The retrieved intensity profile of the optimal pulse consisted from a comb of about eight subpulses (Fig. 6(c)). In contrast, energy in the anti-optimal pulse was mostly concentrated in a single peak pulse of 80 fs (FWHM) in duration (Fig. 6(e)) together with smaller-amplitude subpulses. A Wigner transformation was used to obtain information on instantaneous frequency profile of the pulse shape; the optimal pulse has a main temporal modulation with a period of 145 fs ($230 \pm 14 \, cm^{-1}$), while the anti-optimal pulse is modulated by a period of approximately 215 fs ($155 \, cm^{-1}$). The optimal pulse has a modulation period that is resonant with the low-frequency torsional motions in the excited state. The modes are directly associated with the isomerization process, whereas the anti-optimal pulse is off resonance and out-of-phase with this period. The optimal pulse generates a coherent superposition of the torsional states, *i.e.*, a wavepacket is created on the excited state potential energy surface and is driven from the Franck–Condon state to the final stage, 13-*cis*-retinal *via* a conical intersection by the optimal controlled pulse.

4. Quantum Control of Retinal Isomerization in Rhodopsin

Cis–trans isomerization of retinal in rhodopsin has been actively investigated since the photoreaction is the first step in vision.[14] In this section, a theoretical study of

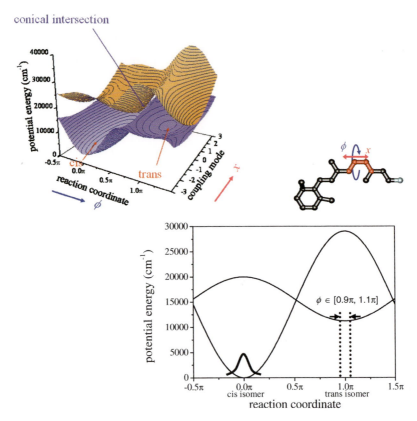

Fig. 7. 2D model potentials in the ground and excited-state S_1 for *cis–trans*-photoisomerization of retinal in rhodopsin. x denotes the coupling coordinate and ϕ denotes the reaction coordinate. The figure at the bottom represents a Gaussian wavepacket located at $\phi = 0$ as the initial *cis*-isomer and the target of the *trans*-isomer located between 0.9π and 1.1π of the reaction coordinates.

quantum control of the *cis–trans* isomerization is presented to clarify the wavepacket dynamics around the conical intersection associated with the two-potential surface crossing.

To simplify the quantum control treatment, a two-dimensional (2D) potential surface model of isomerization of retinal in rhodopsin, which was proposed by Hahn and Stock[15] (Fig. 7), was adopted. The 2D model potential has been shown to reproduce qualitatively the spectroscopic characteristics of retinal in rhodopsin measured by Wang *et al.*[16]

To consider orientation effects of retinal in rhodopsin, it was assumed that there are two orientations, *i.e.*, the transition moments align parallel to a linearly polarized electric field of the control laser field.[17] The density operator of the system is given as

$$\rho(t) = |\Psi_+(t)>\frac{1}{2}<\Psi_+(t)| + |\Psi_-(t)>\frac{1}{2}<\Psi_-(t)|. \quad (1)$$

Here, $|\Psi_+(t)\rangle$ and $|\Psi_-(t)\rangle$, the wavefunctions of retinal molecules with the two orientations, satisfy the time-dependent Schrödinger equation,

$$i\hbar \frac{\partial}{\partial t}|\Psi_\pm(t)\rangle = [H \mp \mu E(t)]|\Psi_\pm(t)\rangle, \tag{2}$$

where H, the molecular Hamiltonian in the diabatic representation, is given as

$$H = -\frac{\hbar^2}{2I}\frac{\partial^2}{\partial \phi^2} - \frac{\hbar^2}{2m}\frac{\partial^2}{\partial x^2} + \begin{bmatrix} V_{11} & V_{12} \\ V_{21} & V_{22} \end{bmatrix}. \tag{3}$$

Here, I and m are the reduced moment of inertia of the *cis–trans* isomerization mode and reduced mass of the coupling mode, respectively. In Eq. (3), V_{kl} with k, $l = 1, 2$ denotes the diabatic potentials. The values of I, m, and V_{kl} were adopted from those by Hahn and Stock.[15] In Eq. (2), μ denotes the electric dipole moment and $E(t)$ denotes an electric field of a linearly polarized laser pulse.

An optimal pulse $E(t)$ is defined to maximize the objective functional J of target operator W at a control time t_f under the constraint of Eq. (2) as

$$J = \frac{1}{2}\langle\Psi_+(t_f)|W|\Psi_+(t_f)\rangle$$
$$+ \frac{1}{2}\langle\Psi_-(t_f)|W|\Psi_-(t_f)\rangle - (\hbar A)^{-1}\int_0^{t_f} dt |E(t)|^2, \tag{4}$$

where positive constant A weighs the physical significance of the penalty.

The optimal pulse is expressed after the variation procedure of Eq. (4) as

$$E(t) = -\frac{A}{2}\text{Im}\{\langle\xi_+(t)|W|\Psi_+(t)\rangle - \langle\xi_-(t)|W|\Psi_-(t)\rangle\}, \tag{5}$$

where $|\xi_+(t)\rangle$ and $|\xi_-(t)\rangle$ are the time-dependent Lagrange multipliers and satisfy

$$i\hbar \frac{\partial}{\partial t}|\xi_\pm(t)\rangle = [H \mp \mu E(t)]|\xi_\pm(t)\rangle, \tag{6}$$

with the final condition $|\xi_\pm(t_f)\rangle = W|\Psi_\pm(t_f)\rangle$.

Figure 8(a) shows the electric field of the optimal pulse for *cis–trans* photoisomerization of retinal, which was calculated by using the optimal control procedure described above, and Fig. 8(b) shows the corresponding time-dependent populations. It can be seen from Fig. 8 that the photoisomerization process can be divided into two stages: the first stage ranges over *ca.* 400 fs from the initial time, in which the wavepacket on the reaction coordinate is shaped by induced electronic transitions. Here, the interval of the pulse sequence corresponds to the frequency of the coupling mode with 1,530 cm^{-1} and the slow modulation component with a period of *ca.* 140 fs originates from the vibrational motion along the reaction coordinates. The other stage ranges from *ca.* 400 fs to the final time 500 fs, in which the shaped wavepacket in the ground state is transferred to the excited state and reaches the target region after passing the conical intersection. A population of 38.8% is

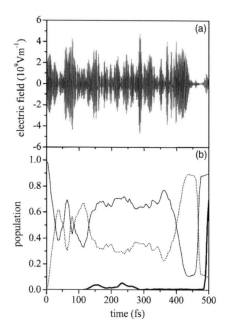

Fig. 8. (a) Optimal control pulse calculated for *cis–trans*-photoisomerization of retinal in rhodopsin. (b) Time evolution of populations in the relevant states in the adiabatic representation: the dotted line denotes the ground-state population, the dotted line denotes the excited-state population, and the solid line denotes the target expectation value. (Reprinted with permission from Abe *et al.*[11] Copyright (2005) by American Institute of Physics).

obtained in the target region by the optimal control procedure, while a population of 16% is obtained by Franck–Condon excitation without considering any control.

To understand the optimized photoisomerization process more clearly, the time-dependent (reduced) population along the reaction coordinate, which is derived by the controlled pulse irradiation, and that derived by the (uncontrolled) Franck–Condon excitation are shown in Fig. 9. The initially localized (uncontrolled) Franck–Condon wavepacket broadens quickly all over the *cis*-isomer region. In contrast to the rapid broadening behavior of the Franck–Condon wavepacket, the controlled wavepacket is localized in the ground electronic state. The shaping subpulses induce multiple electronic transitions to create a vibrational wavepacket that consists of a linear combination of vibrational eigenstates in the ground electronic state. The excitation subpulse transfers *ca.* 70% of the population into the excited state just before the vibrational wavepacket reaches the turning point associated with the narrowest distribution.

It should be noted that to obtain the population in the designated target region as much as possible, the control pulse prepares a wavepacket that is localized along the reaction coordinate with a small amount of energy in the coupling mode. This causes efficient nonadiabatic transition rates at the conical intersection. On the other hand,

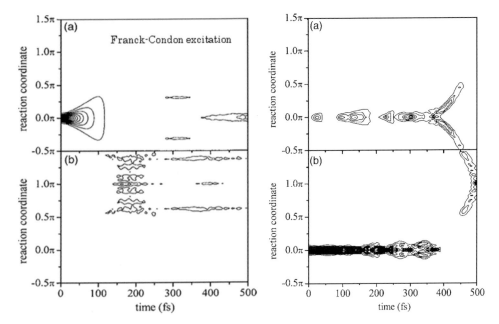

Fig. 9. Time-evolution of the population tracing out the contribution of the coupling mode in the electronic (adiabatic) excited state (a) and that in the (adiabatic) ground state (b); left-hand side for the (uncontrolled) Franck–Condon excitation; right-hand side for the controlled pulse irradiation. (Reprinted with permission from Abe et al.[11] Copyright (2005) by American Institute of Physics).

an uncontrolled excitation pulse creates the initial vibronic distribution, so-called Poisson distribution for both reaction and coupling modes, which is given by the Franck–Condon principle. This causes rapid spreading of the wavepacket over the 2D potential surface in the *cis*-isomer domain, resulting in low photoisomerization yields.

In summary, in this chapter, experimental and theoretical studies on a nuclear quantum dynamics of *cis–trans* isomerization of photoexcited aromatic molecules are presented. Experimental works on the geometrical changes in *cis–trans* isomerization of stilbene and quantum control of retinal in rhodopsin are introduced as typical examples for femtosecond time real observation of nuclear quantum dynamics and quantum control of polyatomic molecules, respectively. Coherent motions of vibrational modes in solution are really observed though they are strongly disturbed by dephasing processes due to solvent–molecule interactions. The ordinary optimal control theory applied to nuclear wavepacket motions on two electronic-state potential surfaces coupled by a conical intersection. Here, a two-mode model was adopted to qualitatively understand the mechanism of the *cis–trans* photoisomerization through a restricted 2D conical intersection. An optimal control treatment in a multi-mode model is required to quantitatively estimate photoisomerization product yields.

References

1. (a) J. Michl and V. Bonačić-Koutecký, *Electronic Aspect of Organic Photochemistry* (Wiley, New York, 1990); (b) G. Orlandi, F. Zerbetto and Z. Zgierski, *Chem. Rev.* **91**, 867 (1991); (c) F. Bernardi, M. Olivucci and M. A. Robb, *Chem. Soc. Rev.* **25**, 321 (1996); M. A. van der Horst and K. J. Hellingwerf, *Acc. Chem. Res.* **37**, 13 (2004).
2. (a) N. Tamai and H. Miyasaka, *Chem. Rev.* **100**, 1875 (2000); (b) S. Hahn and G. Stock, *J. Chem. Phys.* **116**, 1085 (2002).
3. (a) P. Kukura, D. W. McCamant, S. Yoon, D. B. Wandschneider and R. A. Mathies, *Science* **310**, 1006 (2005); (b) C. Dugave, *Cis–Trans Isomerization in Biochemistry* (Wiley-VCH, 2006).
4. (a) T. Yoshizawa and G. Wald, *Nature* **197**, 1279 (1963); (b) R. W. Shoenlein, L. A. Peteanu, R. A. Mathies and C. V. Shank, *Science* **254**, 412 (1991).
5. (a) R. R. Birge, *Annu. Rev. Biophys. Bioeng.* **10**, 315 (1981); R. A. Mathies, C. H. B. Cruz, W. T. Pollard and C. V. Shank, *Science* **240**, 777 (1988).
6. (a) T. Polívka and V. Sundström, *Chem. Rev.* **104**, 2021 (2004); (b) L. Seidner, G. Stock and W. Domcke, *J. Chem. Phys.* **103**, 3998 (1995); (c) D. H. Waldeck, *Chem. Rev.* **91**, 415 (1991); (d) W. von, E. Doering and T. Kitagawa, *J. Am. Chem. Soc.* **113**, 4288 (1991); (e) F. Bernardi, M. Garavelli and M. Olucci IV, *Mol. Phy.* **92**, 359 (1997).
7. (a) M. Klessinger and J. Michl, *Excited States and Photochemistry of Organic Molecules* (Wiley-VCH, New York, 1995); (b) W. Domcke, D. R. Yarkony and H. Köppel, Eds., *Conical Intersections* (World Scientific, Singapore, 2004); (c) B. G. Levine and T. J. Martínez, *Annu. Rev. Phys. Chem.* **58**, 613 (2007).
8. S. Pedersen, L. Bañares and A. H. Zewail, *J. Chem. Phys.* **97**, 8801 (1992).
9. S. Takeuchi, S. Ruhman, T. Tsuneda, M. Chiba, T. Taketsugu and T. Tahara, *Science* **322**, 1073 (2008).
10. V. I. Prokhorenko, A. M. Nagy, S. A. Waschuk, L. S. Brown, R. R. Birge and R. J. Dwayne Miller, *Science* **313**, 1257 (2006).
11. M. Abe, Y. Ohtsuki, Y. Fujimura and W. Domcke, *J. Chem. Phys.* **123**, 144508 (2005).
12. D. C. Todd and G. R. Fleming, *J. Chem. Phys.* **98**, 269 (1993).
13. (a) T. Kobayashi, T. Saito and H. Ohtani, *Nature* **414**, 531 (2001); (b) T. Ye, N. Friedman, Y. Gat, G. H. Atkinson, M. Sheves, M. Ottolenghi and S. Ruhman, *J. Phys. Chem. B* **103**, 5122 (1999); (c) M. Nonella, *J. Phys. Chem. B* **104**, 11379 (2000); (d) S. Hayashi, E. Tajkhorshid and K. Schulten, *Biophys. J.* **83**, 1281 (2002); (e) S. Schenki, F. van Mourik, N. Freidman, M. Sheves, R. Schlesinger, S. Haacke and M. Chergui, *PNAS* **103**, 4101 (2006).
14. (a) L. A. Peteanu, R. W. Schoenlein, Q. Wang, R. A. Mathies and C. A. Shank, *PNAS* **90**, 11762 (1993); (b) H. Chosrowjan, N. Mataga, Y. Shibata, S. Tachibanaki, H. Kandori, Y. Shichida, T. Okada and T. Kouyama, *J. Am. Chem. Soc.* **120**, 9706 (1998); (c) G. Haran, E. A. Morlino, J. Matthes, R. H. Callender and R. M. Hochstrasser, *J. Phys. Chem. A* **103**, 2202 (1999); (c) M. Garavelli, F. Negri and M. Olivucci, *J. Am. Chem. Soc.* **121**, 1023 (1999); (d) L. D. Vico, C. S. Page, M. Garavelli, F. Bernardi, R. Basosi and M. Olivucci, *J. Am. Chem. Soc.* **124**, 4124 (2002); S. C. Flores and V. S. Batista, *J. Phys. Chem. B* **108**, 6745 (2004); (e) L. M. Frutos, T. Andruniów, F. Santoro, N. Ferré and M. Olivucci, *PNAS* **104**, 7764 (2007).

15. S. Hahn and G. Stock, *J. Phys. Chem. B* **104**, 1146 (2000).
16. Q. Wang, R. W. Schoenlein, L. A. Peteanu, R. A. Mathies and C. V. Shank, *Science* **266**, 422 (1994).
17. (a) M. Machholm and N. E. Henriksen, *J. Chem. Phys.* **111**, 3051 (1999); (b) K. Nakagami, Y. Ohtsuki and Y. Fujimura, *J. Chem. Phys.* **117**, 6429 (2002).

Chapter 8

Quantum Control of Molecular Chirality

In this chapter, several control scenarios of molecular chirality transformation by using electric fields of linearly polarized lasers are presented from the theoretical viewpoint. First, chirality transformation of preoriented enantiomers in a racemic mixture is treated. The scenarios involve pump–dump control, stimulated Raman adiabatic passage, and a quantum control method. The control scenarios are applied to both enantiomers with axial chirality and those with helical chirality. Chirality transformation of randomly oriented enantiomers by three polarization components of the electric field of the laser is finally presented.

1. Molecular Chirality Transformation in a Preoriented Racemic Mixture

It is well known that circularly polarized electric fields of visible or UV lights can predominantly decompose a one-enantiomeric constituent of chiral molecules in a racemic mixture.[1] The decomposition, which is called asymmetric destruction or asymmetric photolysis, utilizes the difference in molecular absorption coefficients between the two-enantiomeric constituents. The perturbations inducing the difference is a second-order interaction involving electric and magnetic dipole moments. Occurrence of intense and circularly polarized electric fields is one of the possible mechanisms of homochirality, *i.e.*, an almost one-enantiomeric constituent of the chiral molecules, in life in earth.[2] It is, however, insufficient to produce an appreciable enantiomeric excess using circularly polarized lasers in ordinary experimental conditions. This is because the magnetic field interaction is very weak in comparison to the electric dipole interaction. In addition, the asymmetric destruction described above is not an effective method for creation of one of the pure enantiomers because fragment species are produced from unwanted enantiomers at the same time. Therefore, it is desirable to produce one of the two enantiomeric constituents, say L- or R-enantiomers, in a racemic mixture by utilizing properties of only electric field component of lights.

Several ideas for molecular chirality transformation by using the coherent properties of lasers have been proposed.[3] In this section, a basic principle of molecular chirality transformation in a racemic mixture by using a nonhelical and linearly polarized electric field of lasers is presented. It should be noted that enantiomers oriented at a surface or in a space can be transformed by noting the photon polarization direction. Orientation of molecules by a laser technique has already been discussed in Chap. 4. Since a pair of preoriented enantiomers has a mirror image with each other, there is a difference in the direction of their transition dipole moments between L- and R-enantiomers, although their magnitudes are equal in a good approximation.

Let us consider a simple model in which the reaction potential is given by a one-dimensional (1D) double-well in the ground state (Fig. 1) as shown in Sec. 11 of Chap. 1. This system consists of four states, $|0_+\rangle$, $|0_-\rangle$, $|1_+\rangle$, and $|1_-\rangle$. Here $|0_+\rangle$ and $|0_-\rangle$ are nearly degenerated with energy ε_0 because of a high potential barrier between L and R.

Let the energy difference between $|0_+\rangle$ ($|0_-\rangle$) and $|1_+\rangle$ ($|1_-\rangle$) be much larger than kT. In this case, the initial density operator $\rho(t_0)$ of the racemic mixture at $t = t_0$ can be expressed in terms of the eigenstates as

$$\rho(t_0) = |0_+\rangle \frac{1}{2} \langle 0_+| + |0_-\rangle \frac{1}{2} \langle 0_-|. \tag{1}$$

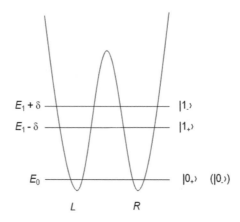

Fig. 1. A four-state model for a molecular chirality transformation. L (R) denotes the L-(R-) enantiomer. The lowest vibrational state $|0_+\rangle$ and the first excited state $|0_-\rangle$ are nearly degenerated with energy E_0. The second excited state with energy $E_1 - \delta$ is denoted by $|1_+\rangle$ and the third excited state with $E_1 + \delta$ is denoted by $|1_-\rangle$. Here, 2δ is the energy separation between the two excited states.

The initial density operator, Eq. (1), can also be expressed in terms of the localized states in the two potential wells, $|v_R\rangle$ and $|v_L\rangle$ ($v = 0$ or 1)

$$|v_R\rangle = \frac{1}{\sqrt{2}}(|v_+\rangle - |v_-\rangle), \tag{2a}$$

and

$$|v_L\rangle = \frac{1}{\sqrt{2}}(|v_+\rangle + |v_-\rangle) \tag{2b}$$

as

$$\rho(t_0) = |0_R\rangle \frac{1}{2} \langle 0_R| + |0_L\rangle \frac{1}{2} \langle 0_L|. \tag{3}$$

This expresses the racemic mixture in which L- and R-enantiomers have equal populations at the initial time.

The key for chirality transformation in a preoriented racemic mixture is how to localize a wavepacket in two states in a one-to-one statistical mixture with magnitude of 1/2. It is well known that the eigenvalues of the system density operator $\rho(t)$ represent statistical weight and are invariant when the time-evolution of $\rho(t)$ is driven by a unitary operator. Therefore, for the initial density operator given by Eqs. (1) and (3) and dipole interaction between lasers and enantiomers, the final state must also be described in terms of a one-to-one statistical mixture.[4] For example, for obtaining the R-enantiomers from a racemic mixture, the target operator $\hat{W}$ is expressed as

$$W = |0_R\rangle \frac{1}{2} \langle 0_R| + |1_R\rangle \frac{1}{2} \langle 1_R|. \tag{4}$$

The Hamiltonian of the total system, $H(t)$, is expressed within the semiclassical treatment of the radiation interaction with matter as

$$H(t) = H_0 - \mu(t). \tag{5}$$

Here, H_0 is the molecular Hamiltonian, μ is the dipole moment vector, and $\boldsymbol{E}(t)$ is the electric field of the laser:

$$\boldsymbol{E}(t) = 2\boldsymbol{A}(t) \cos \omega t, \tag{6}$$

where $\boldsymbol{A}(t)$ is the pulse envelope function with a photon-polarization vector $\boldsymbol{A}$ and ω is the carrier frequency of the laser.

The Hamiltonian matrix, $\boldsymbol{H}(t)$, can be expressed in the eigenstate representation as

$$\boldsymbol{H}(t) = \begin{pmatrix} \varepsilon_0 & 0 & -\langle 0_+|\boldsymbol{\mu}|1_+\rangle \cdot \boldsymbol{E}(t) & -\langle 0_+|\boldsymbol{\mu}|1_-\rangle \cdot \boldsymbol{E}(t) \\ 0 & \varepsilon_0 & -\langle 0_-|\boldsymbol{\mu}|1_+\rangle \cdot \boldsymbol{E}(t) & -\langle 0_-|\boldsymbol{\mu}|1_-\rangle \cdot \boldsymbol{E}(t) \\ -\langle 1_+|\boldsymbol{\mu}|0_+\rangle \cdot \boldsymbol{E}(t) & -\langle 1_+|\boldsymbol{\mu}|0_-\rangle \cdot \boldsymbol{E}(t) & \varepsilon_1 - \delta & 0 \\ -\langle 1_-|\boldsymbol{\mu}|0_+\rangle \cdot \boldsymbol{E}(t) & -\langle 1_-|\boldsymbol{\mu}|0_-\rangle \cdot \boldsymbol{E}(t) & 0 & \varepsilon_1 + \delta \end{pmatrix}. \tag{7}$$

The matrix elements in the localized basis set can be expressed in a good approximation as

$$\langle 1_L|\boldsymbol{\mu}|0_R\rangle = 0 \quad \text{and} \quad \langle 1_R|\boldsymbol{\mu}|0_L\rangle = 0. \tag{8}$$

This leads to the following relation between the dipole matrix elements in the eigenstate representation:

$$\langle 1_+|\boldsymbol{\mu}|0_+\rangle = \langle 1_-|\boldsymbol{\mu}|0_-\rangle \quad \text{and} \quad \langle 1_+|\boldsymbol{\mu}|0_-\rangle = \langle 1_-|\boldsymbol{\mu}|0_+\rangle. \tag{9}$$

Selective preparation of preoriented enantiomers depends on the photon polarization direction of the laser field for a pumping process. For an L- to R-enantiomer transformation, L-enantiomers are optically active between the ground state $|0_L\rangle$ and excited states $|1_\pm\rangle$, while R-enantiomers are inactive if the polarization of the laser is set as

$$-\langle 1_+|\boldsymbol{\mu}|0_+\rangle \cdot \boldsymbol{A}(t) = -\langle 1_+|\boldsymbol{\mu}|0_-\rangle \cdot \boldsymbol{A}(t) \equiv \hbar\Omega(t). \tag{10}$$

The above expression is equivalent to

$$\langle 1_+|\boldsymbol{\mu}|0_R\rangle \cdot \boldsymbol{A}(t) = 0 \tag{11a}$$

and

$$\langle 1_+|\boldsymbol{\mu}|0_L\rangle \cdot \boldsymbol{A}(t) = -\sqrt{2}\hbar\Omega(t). \tag{11b}$$

In a similar way, from Eqs. (8) and (10), the following expression can be derived

$$\langle 1_-|\boldsymbol{\mu}|0_R\rangle \cdot \boldsymbol{A}(t) = 0 \quad \text{and} \quad \langle 1_-|\boldsymbol{\mu}|0_L\rangle \cdot \boldsymbol{A}(t) = -\sqrt{2}\hbar\Omega(t). \tag{12}$$

Equations (11) and (12) are equivalent to

$$\langle 1_R|\boldsymbol{\mu}|0_R\rangle \cdot \boldsymbol{A}(t) = 0 \tag{13a}$$

and

$$\langle 1_L|\boldsymbol{\mu}|0_L\rangle \cdot \boldsymbol{A}(t) = -2\hbar\Omega(t), \quad \text{respectively,} \tag{13b}$$

as shown in Fig. 2.

If an electric field of a linearly polarized laser whose polarization direction is determined by Eq. (13a), the population of R-enantiomers, $P_R(t)$, is expressed as

$$P_R(t) = \langle 0_R|\rho(t)|0_R\rangle + \langle 1_R|\rho(t)|1_R\rangle \geq \frac{1}{2}. \tag{14}$$

The population of R-enantiomers in the excited state $|1_R\rangle$, $\langle 1_R|\rho(t)|1_R\rangle$, is transferred from L-enantiomers in the ground state $\langle 0_L|\rho(t_0)|0_L\rangle = 1/2$ via a tunneling process after optical excitation. Thus, the total population of R-enantiomers, $P_R(t)$, is increased compared with that in the initial racemic mixture.

Similarly, a chirality transformation from an R-enantiomer to an L-enantiomer in a preoriented racemic mixture can be performed by setting the laser field

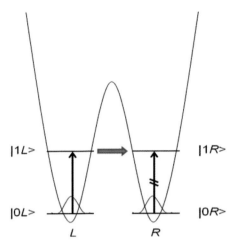

Fig. 2. Selective excitation of L-enantiomers for chiral transformation from L- to R-enantiomers in a racemic mixture. The bold arrow denotes tunneling from the potential well of the L-enantiomer to that of the R-one.

polarization as

$$-\langle 1_+|\boldsymbol{\mu}|0_+\rangle \cdot \boldsymbol{A}'(t) = \langle 1_+|\boldsymbol{\mu}|0_-\rangle \cdot \boldsymbol{A}'(t) \equiv \hbar\Omega'(t). \tag{15}$$

Then, we obtain

$$\langle 1_L|\boldsymbol{\mu}|0_L\rangle \cdot \boldsymbol{A}'(t) = 0 \quad \text{and} \quad \langle 1_R|\boldsymbol{\mu}|0_R\rangle \cdot \boldsymbol{A}'(t) = -2\hbar\Omega'(t). \tag{16}$$

This is the basic principle for laser control of enantiomers in a preoriented racemic mixture by taking into account the photon polarization direction of the pump laser.

Let us now begin with a quantitative treatment of chirality transformation in the racemic mixture. An analytical expression for the time-dependent populations of the enantiomers is derived in the dressed state representation.[5] The time evolution of the density operator $\hat{\rho}(t)$ obeys the Liouville equation

$$i\hbar \frac{d}{dt}\rho(t) = [H(t), \rho(t)], \tag{17}$$

where [,] is a commutator.

For a stationary laser field, the equation of motion of the density matrix $\rho(t)$ represented by field-free eigenstates is

$$i\hbar \frac{d}{dt}\rho(t) = \left[\begin{pmatrix} \varepsilon_0 & 0 & 2\hbar\Omega\cos(\omega t) & 2\hbar\Omega\cos(\omega t) \\ 0 & \varepsilon_0 & 2\hbar\Omega\cos(\omega t) & 2\hbar\Omega\cos(\omega t) \\ 2\hbar\Omega\cos(\omega t) & 2\hbar\Omega\cos(\omega t) & \varepsilon_1 - \delta & 0 \\ 2\hbar\Omega\cos(\omega t) & 2\hbar\Omega\cos(\omega t) & 0 & \varepsilon_1 + \delta \end{pmatrix}, \rho(t) \right] \tag{18}$$

with the initial condition

$$\rho(t_0) = \begin{pmatrix} 1/2 & 0 & 0 & 0 \\ 0 & 1/2 & 0 & 0 \\ 0 & 0 & 0 & 0 \\ 0 & 0 & 0 & 0 \end{pmatrix}. \quad (19)$$

In Eq. (18), Rabi frequency, Ω, is defined as $\hbar\Omega = -\langle 1_+|\boldsymbol{\mu}|0_+\rangle \cdot \boldsymbol{A}$.

Equation (18) can be solved in the rotating wave approximation. The time-dependent population in each localized state can be expressed as[5]

$$P_{0R} = \langle 0_R|\rho(t)|0_R\rangle = \frac{1}{2}, \quad (20a)$$

$$P_{0L} = \langle 0_L|\rho(t)|0_L\rangle = \frac{1}{2}\left[\frac{\delta^2 + 4\hbar^2\Omega^2\cos(\tilde{\Omega}t)}{\hbar^2\tilde{\Omega}^2}\right]^2, \quad (20b)$$

$$P_{1R} = \langle 1_R|\rho(t)|1_R\rangle = 2\left[\frac{\delta\Omega[1-\cos(\tilde{\Omega}t)]}{\hbar\tilde{\Omega}^2}\right]^2, \quad (20c)$$

and

$$P_{1L} = \langle 1_L|\rho(t)|1_L\rangle = 2\left[\frac{\Omega\sin(\tilde{\Omega}t)}{\tilde{\Omega}}\right]^2. \quad (20d)$$

with $\tilde{\Omega} = \sqrt{\frac{\delta^2}{\hbar^2} + 4\Omega^2}$.

Equation (20d) indicates that the population in $|1_L\rangle$ is maximized at $\tilde{\Omega}t = \pi/2$ and the population of $|1_R\rangle$ transferred by tunneling from $|1_L\rangle$ is maximized at $\tilde{\Omega}t = \pi$. The maximum population in the localized state, $|1_R\rangle$, is obtained if Rabi frequency satisfies $|\Omega| = \delta/(2\hbar)$. The total population in the right-hand well is expressed as

$$P_R = \frac{1}{2} + 2\left[\frac{\Omega\sin(\tilde{\Omega}t)}{\tilde{\Omega}}\right]^2. \quad (21)$$

We now take a phosphinotioic acid, H_2POSH, with axial chirality as a chiral molecule to demonstrate how enantiomers in a preoriented racemic mixture are selectively prepared by applying a linearly polarized laser. The reaction path ϕ is along the torsional coordinate of the SH bond around the PS bond of H_2POSH. Let H_2POSH be preoriented in such a way that the PS bond is fixed in the $X-Z$ plane. The potential is characterized by a double-well potential as shown in Fig. 1.

An *ab initio* molecular orbital calculation indicates that the L-form enantiomer is located at a torsional angle of $-60°$, and the R-form is located at $60°$. Its barrier height is 494 cm^{-1}.[3(e)] It is reasonable to adopt a four-state model consisting of the ground, first, second, and third excited states. The condition of quasi-degenerate between the ground and first excited states is satisfied within a time scale of a few hundred ps since level splitting between them is only 0.053 cm^{-1}. This corresponds to tunneling time of 630 ps. The level splitting between the second and third excited states, 2δ, is 1.6 cm^{-1}, which corresponds to tunneling time of 21 ps. The dipole moments μ_X and μ_Y, which are symmetric and antisymmetric, respectively, with respect to the ZX-plane involving O, P, and S atoms, were taken into account in determining the direction of the photon polarization. From *ab initio* MO results of $|\langle 1_+|\mu|0_L\rangle| = 0.19$ Debye, the direction of the photon polarization with respect to the X-axis was estimated to be 58°. It should be noted that the polarization direction is almost parallel to the SH-bond of H$_2$POSH in the R-form as shown in Fig. 3. A laser field with the polarization direction makes chirality transformation from L- to R-enantiomers possible.

Figure 4 shows the time-evolution of the chirality transformation in a racemic mixture under an optimal condition. Here, Rabi frequency satisfies the condition

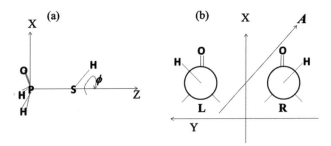

Fig. 3. (a) Geometrical structure of H$_2$POSH with axitial chirality and (b) L- and R-enantiomers of H$_2$POSH and photon polarization direction (thick arrow).

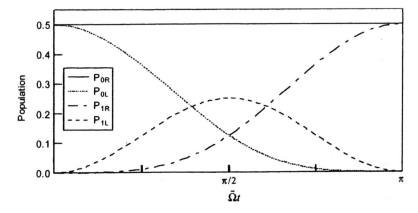

Fig. 4. Time-evolution of the chirality transformation under an optimal condition. (Reprinted with permission from Hoki and Fujimura.[5] Copyright (2002) by American Chemical Society).

$|\Omega| = \delta/(2\hbar)$, under which an R-enantiomer of 100% is produced from the racemic mixture. For this ideal case, both the Rabi oscillation and tunneling rate are synchronously matched with each other. Therefore, all of the population of $|1_L\rangle$ created from the ground state is transferred to the target state, $|1_R\rangle$, by a tunneling process.

2. Pump–Dump Control *via* an Electronic Excited State

Let us take a chirality transformation *via* an electronic excited state. In this scenario, the control event is carried out using pump and dump pulses in a femtosecond time scale.[6] Since a pair of preoriented enantiomers is mirror images, there is a difference in the direction of their transition dipole moments between L- and R-enantiomers, i.e., $\langle e_m|\boldsymbol{\mu}|g_L\rangle \neq \langle e_m|\boldsymbol{\mu}|g_R\rangle$, although their magnitudes are equal. Here, $|g_L\rangle(|g_R\rangle)$ denotes the electronic ground state of L-(R-)enantiomers and $|e_m\rangle$ denotes the electronic excited state.

Consider a selective preparation of L-enantiomers from their racemic mixture. For simplicity, the chiral molecule is assumed to be characterized by a 1D double-well potential in the electronic ground state and by a single-well potential in an electronic excited state as shown in Fig. 5. In the localized basis set, the degenerate lowest vibrational states with energy ε_{g0} in the ground electronic state are denoted by $|g0_R\rangle$ and $|g0_L\rangle$, and degenerate nth vibrationally excited states with energy ε_{gn} are denoted by $|gn_R\rangle$ and $|gn_L\rangle$. The mth vibronic state is denoted by $|em\rangle$.

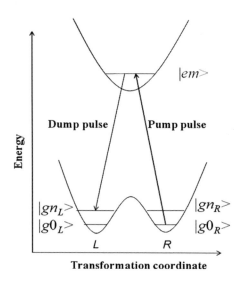

Fig. 5. A simple model for a pump–dump control *via* an electronic excited state.

In a low temperature limit, the initial density operator $\rho(t_0)$ is expressed as

$$\rho(t_0) = |g0_R\rangle \frac{1}{2} \langle g0_R| + |g0_L\rangle \frac{1}{2} \langle g0_L|. \tag{22}$$

We now specify the target operator that produces L-enantiomers from a racemic mixture. The eigenvalues of $\rho(t)$ represent statistical weight and are invariant when the time evolution of $\rho(t)$ is driven by a unitary operator. The final state must also be described in terms of a one-to-one statistical mixture.[4] Therefore, the target operator for L-enantiomers is given as

$$|g0_L\rangle \frac{1}{2} \langle g0_L| + |gn_L\rangle \frac{1}{2} \langle gn_L|.$$

The semi-classical Hamiltonian, $H(t)$, is expressed as

$$H(t) = H_0 - \boldsymbol{\mu} \cdot \boldsymbol{E}(t). \tag{23}$$

Here, H_0 is the molecular Hamiltonian, $\boldsymbol{\mu}$ is the dipole moment, and $\boldsymbol{E}(t)$ is the electric field for pump and dump laser pulses:

$$\boldsymbol{E}(t) = 2\boldsymbol{f}_p(t)\cos(\omega_p t) + 2\boldsymbol{f}_d(t)\cos(\omega_d t), \tag{24}$$

where $\boldsymbol{f}_p(t)$ ($\boldsymbol{f}_d(t)$) denotes an envelope function of the pump (dump) pulse with a central frequency ω_p (ω_d) and with a photon polarization vector. The time evolution of the molecular system, $\rho(t)$, is determined by solving the Liouville equation:

$$i\hbar \frac{d}{dt}\rho(t) = [H(t), \rho(t)]. \tag{25}$$

To derive an analytical expression for the transfer yield of molecular handedness, the two pulses are assumed to be separated temporarily from each other. The time evolution of the molecular system can be divided into two independent processes: one is the pump process in which the population is created in the excited vibronic state by the pump pulse and the other is the dump process in which the population is transferred from the excited state to the final target state by the dump pulse.

Let us consider the pumping process by which the population of R-enantiomers is transferred from the initial ground state to the excited vibronic state, while the population of L-enantiomers in the ground state remains unchanged. The pump pulse satisfies two conditions:

$$\langle em|\boldsymbol{\mu}|g0_L\rangle \cdot \boldsymbol{f}_p(t) = 0 \tag{26a}$$

and

$$\langle em|\boldsymbol{\mu}|g0_R\rangle \cdot \boldsymbol{f}_p(t) = -\hbar\Omega_p(t). \tag{26b}$$

Here, $|\Omega_p(t)|$ is the so-called Rabi frequency. Equation (26a) indicates that the polarization direction of the pump pulse is taken to be orthogonal to the direction of the transition moment between the electronic ground and excited states.

The time evolution of the molecular system in the pump process is obtained by solving the equation of motion of the density matrix $\rho(t)$ expressed in terms of $|g0_R\rangle$, $|g0_L\rangle$ and $|em\rangle$ as

$$i\hbar\frac{d}{dt}\rho(t) = \left[\begin{pmatrix} \varepsilon_{g0} & 0 & 0 \\ 0 & \varepsilon_{g0} & 2\hbar\Omega_p(t)\cos(\omega_p t) \\ 0 & 2\hbar\Omega_p(t)\cos(\omega_p t) & \varepsilon_{em} \end{pmatrix}, \rho(t)\right] \tag{27}$$

with the initial condition

$$\rho(t_0) = \begin{pmatrix} 1/2 & 0 & 0 \\ 0 & 1/2 & 0 \\ 0 & 0 & 0 \end{pmatrix}. \tag{28}$$

The central frequency of the pump pulse is set as $\hbar\omega_p = \varepsilon_{em} - \varepsilon_{g0}$ under the resonant condition. Within the rotating wave approximation, Eq. (27) is rewritten as

$$i\hbar\frac{d}{dt}\tilde{\rho}(t) = \left[\begin{pmatrix} 0 & 0 & 0 \\ 0 & 0 & \hbar\Omega_p(t) \\ 0 & \hbar\Omega_p(t) & 0 \end{pmatrix}, \tilde{\rho}(t)\right], \tag{29}$$

where

$$\tilde{\rho}(t) = \mathbf{R}(t)\rho(t)\mathbf{R}^{-1}(t) \tag{30}$$

with

$$\mathbf{R}(t) = \begin{pmatrix} \exp\left\{\frac{i\varepsilon_{g0}(t-t_0)}{\hbar}\right\} & 0 & 0 \\ 0 & \exp\left\{\frac{i\varepsilon_{g0}(t-t_0)}{\hbar}\right\} & 0 \\ 0 & 0 & \exp\left\{\frac{i\varepsilon_{em}(t-t_0)}{\hbar}\right\} \end{pmatrix}. \tag{31}$$

The solution of the equation of motion, Eq. (29), is given as

$$\rho(t) = \mathbf{R}^{-1}(t)\mathbf{U}(t,t_0)\tilde{\rho}(t_0)\mathbf{U}(t_0,t)\mathbf{R}(t) \tag{32}$$

with $\tilde{\rho}(t_0) = \rho(t_0)$.

Here,

$$\mathbf{U}(t, t_0) = \begin{pmatrix} 1 & 0 & 0 \\ 0 & \cos\dfrac{\phi_p(t, t_0)}{2} & -i\sin\dfrac{\phi_p(t, t_0)}{2} \\ 0 & -i\sin\dfrac{\phi_p(t, t_0)}{2} & \cos\dfrac{\phi_p(t, t_0)}{2} \end{pmatrix}, \quad (33)$$

where ϕ_p is the pump–pulse area defined as

$$\phi_p(t, t_0) = 2\int_{t_0}^{t} \Omega_p(t')dt'. \quad (34)$$

The diagonal density matrix elements are finally expressed in an analytical form as

$$\rho_{g0_L, g0_L}(t) = \frac{1}{2}, \quad (35a)$$

$$\rho_{g0_R, g0_R}(t) = \frac{1}{2}\cos^2\frac{\phi_p(t, t_0)}{2} \quad (35b)$$

and

$$\rho_{em, em}(t) = \frac{1}{2}\sin^2\frac{\phi_p(t, t_0)}{2}. \quad (35c)$$

It can be seen from Eq. (35) that the population is completely transferred to the vibronic state $|em\rangle$ from the lowest vibrational ground state $|g0_R\rangle$ for $\phi_p(t, t_0) = \pi$, i.e., by using a π pump pulse.

Next, consider the dump process to create L-enantiomers in the vibrationally excited ground state $|gn_L\rangle$ from $|em\rangle$ by a laser pulse. The initial population in the dump process is given by Eq. (35c). The polarization vector of the pump pulse is determined by the conditions

$$\langle gn_R|\boldsymbol{\mu}|em\rangle \cdot \boldsymbol{f}_d(t) = 0 \quad \text{and} \quad \langle gn_L|\boldsymbol{\mu}|em\rangle \cdot \boldsymbol{f}_d(t) = -\hbar\Omega_d(t). \quad (36)$$

Here, $|\Omega_d(t)|$ is the magnitude of Rabi frequency in the dump process.

By using the same treatment as that described for the pump process, we obtain an analytical expression for the population in the target state $|gn_L\rangle$ and the vibronic state at t_f. The transfer yield of L-enantiomers $Y_L(t_f)$ can be expressed as

$$Y_L(t_f) = \langle g0_L|\rho(t_f)|g0_L\rangle + \langle gn_L|\rho(t_f)|gn_L\rangle$$
$$= \frac{1}{2}\left\{1 + \sin^2\frac{\phi_d(t_f, t)}{2}\sin^2\frac{\phi_p(t, t_0)}{2}\right\}. \quad (37)$$

Here, $\phi_d(t_f, t) = 2\int_t^{t_f} \Omega_d(\tau)d\tau$. The maximum population is obtained by applying both π pump and π dump pulses to the racemic mixture.

So far, a laser control scheme to create L-handed enantiomers from a racemic mixture has been described. In an analogous way, it is possible to create R-handed enantiomers from a racemic mixture.

Let us take H_2POSH again as a real chiral molecule to demonstrate the effectiveness of the pump–dump pulse method. Both the electronic ground and singlet excited state potential energy curves and transition dipole moments have been calculated using time-dependent density functional theory.[7] L-enantiomers are prepared from a racemic mixture by setting $|g0_L\rangle \frac{1}{2} \langle g0_L| + |g1_L\rangle \frac{1}{2} \langle g1_L|$ as the target operator. The tunneling time defined as the whole oscillation period R→L→R in the first vibrationally excited doublet is estimated to be 34 ps.[7] This means that the control has to be completed within this time. The vibronic state of fifth torsional quantum number $m = 5$, $|e5\rangle$ in the first singlet excited state was adopted as the intermediate state because of the main transition moment from the lowest ground state.[7]

The direction of the polarization vector of laser propagating along the X-axis was 45° with respect to the P–S bond of H_2POSH for the pumping process from Eqs. (26) and (36). Then, the polarization vector with the envelope function $f_p(t)$ of the pump pulse was set as

$$f_p(t) = \begin{cases} (e_Y + e_Z) E_p \sin^2(\alpha_p t) & (\text{for } 0 \leq t \leq 300 \text{ fs}) \\ 0 & (\text{for } 300 \text{ fs} < t) \end{cases}, \quad (38a)$$

with $E_p = 7.4 \times 10^9$ V/m and $\alpha_p = \pi/300 \text{ fs}^{-1}$. Here, e_Y and e_Z are the unit vectors of Y and Z directions in the space-fixed coordinates, respectively.

In a similar way, the direction of the polarization vector of the dump pulse is −45° to the P–S bond. The polarization vector with the envelope function $f_d(t)$ of the dump pulse was set as

$$f_p(t) = \begin{cases} (e_Y - e_Z) E_d \sin^2\{\alpha_d(t - 300)\} & (\text{for } 300 \leq t \leq 900 \text{ fs}) \\ 0 & (\text{for } t < 300 \text{ fs}, 900 \text{ fs} < t) \end{cases},$$

(38b)

with $E_d = 2.2 \times 10^9$ V/m and $\alpha_d = \pi/600 \text{ fs}^{-1}$.

The bandwidth of the pump pulse and that of the dump pulse were estimated to be 50 and 25 cm^{-1}, respectively, by using Fourier transformation of Eqs. (38a) and (38b). These band widths are small compared with the level spacings of ∼200 cm^{-1} in the ground state and those of ∼300 cm^{-1} in the resonant intermediate electronic state. Therefore, the pump and dump control method can be applied almost perfectly within the approximate five-state model.

Figures 6(a) and 6(b) show the time evolution of the wavepackets in the lowest torsional state and that in the first excited quantum state in the electronic ground state, respectively. It can be seen from Fig. 6 how effectively the molecular chirality is controlled by the pump and dump control method.

Figure 7 shows temporal behaviors of the population of localized states for the pump–dump control scheme. It can be seen that the population of $|g0_R\rangle$ is transferred to $|g1_L\rangle$ through $|e5\rangle$ with a high yield of $Y_L(t_f) = 0.96$ within $t_f = 800$ fs.

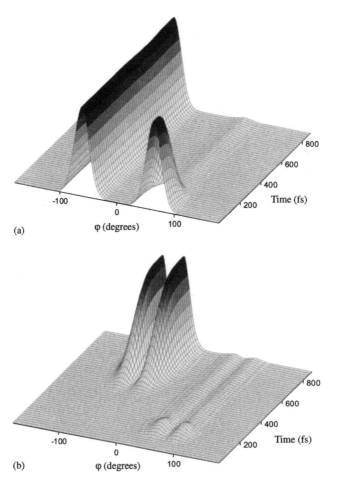

Fig. 6. Time evolution of wavepackets in chiral transformation in pump–dump pulse control. (a) Wavepackets of $|g0_R\rangle$ and $|g0_L\rangle$ and (b) wavepackets of $|g1_R\rangle$ and $|g1_L\rangle$. (Reprinted with permission from Hoki et al.[6] Copyright (2002) by American Institute of Physics).

The present results will serve as a reference for extended studies including additional degrees of freedom and competing processes such as photodissociations.[8]

3. Stimulated Raman Adiabatic Passage Method

Stimulated Raman adiabatic passage (STIRAP) is one of the methods for complete population transfer.[9] This method is carried out by applying the Stokes pulse before the pump laser counter-intuitively. The characteristics of STIRAP are that it is robust and requires no sophisticated experimental setup. STIRAP has been applied to atomic or molecular systems with different energies between the initial and final states. This was necessary for the pump and Stokes pulses to be distinguished.

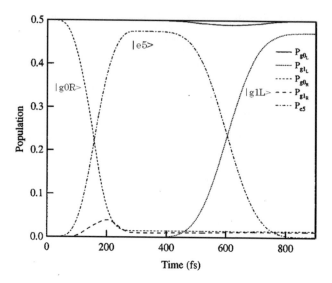

Fig. 7. Time evolution of populations of H_2POSH. The solid, dotted, broken, thick-broken, and dotted-broken lines denote the populations of $|g0_L\rangle$, $|g1_L\rangle$, $|g0_R\rangle$, $|g1_R\rangle$, and $|e5\rangle$, respectively. (Reprinted with permission from Hoki et al.[6] Copyright (2002) by American Institute of Physics).

In other words, the original STIRAP cannot be applied to molecular chirality transformation since the initial and final states have the same energy. In this subsection, the outline of a new version of STIRAP,[10] which is applicable to a chirality transformation, is presented.

We take a population transfer from the initial state $|0_L\rangle$ to a final state $|0_R\rangle$ via an intermediate state $|em\rangle$. The electric field of STIRAP is expressed as

$$\boldsymbol{E}(t) = \boldsymbol{e}_1 A_1(t) \cos(\omega_1 t) + \boldsymbol{e}_2 A_2(t) \cos(\omega_2 t), \tag{39}$$

where $\boldsymbol{e}_1$ ($\boldsymbol{e}_2$) is the unit vector of polarization of photons with central frequency ω_1 (ω_2). Both frequencies are assumed to be resonant to the transitions between $|0_L\rangle$ ($|0_R\rangle$) and $|em\rangle$ and are simplified as $\omega_1 = \omega_2 = \omega$. $A_1(t)$ ($A_2(t)$) denotes the envelope of laser pulse 1(2).

Equation (39) can be rewritten as

$$\boldsymbol{E}(t) = \frac{\boldsymbol{A}(t)}{|\boldsymbol{A}(t)|} |\boldsymbol{A}(t)| \cos(\omega t) = \boldsymbol{\eta}(t) A(t) \cos(\omega t), \tag{40}$$

where $\boldsymbol{A}(t) = \boldsymbol{e}_1 A_1(t) + \boldsymbol{e}_2 A_2(t)$, $A(t) \equiv |\boldsymbol{A}(t)|$, and

$$\boldsymbol{\eta}(t) \equiv \frac{\boldsymbol{A}(t)}{|\boldsymbol{A}(t)|}. \tag{41}$$

Equation (40) indicates that two lasers with the same frequency and two different linear polarizations are equivalent to a single laser with a time-dependent polarization vector.

Chiral molecular Hamiltonian in the interaction picture can generally be expressed within the rotating wave approximation (RWA) under a resonance condition as

$$H_I(t) = -\frac{\eta(t)A(t)}{2} \cdot (|0_L\rangle \boldsymbol{\mu}_{0L,em}\langle em| + |0_R\rangle \boldsymbol{\mu}_{0R,em}\langle em|) + \text{h.c.}, \quad (42)$$

where $\boldsymbol{\mu}_{0L,em}$ ($\boldsymbol{\mu}_{0R,em}$) is the transition dipole moment in the L-(R-)enantiomer. Equation (42) can be rewritten as

$$H_I(t) = -\frac{A(t)}{2}\mu_{0,em}(\cos\eta_L(t)|0_L\rangle\langle em| + \cos\eta_R(t)|0_L\rangle\langle em|) + \text{h.c.} \quad (43)$$

Here, $\eta_L(t)$ ($\eta_R(t)$) is the relative angle between $\boldsymbol{\eta}(t)$ and $\boldsymbol{\mu}_{0L,em}$ ($\boldsymbol{\mu}_{0L,em}$), and $\mu_{0,em} \equiv |\boldsymbol{\mu}_{0L,em}| = |\boldsymbol{\mu}_{0R,em}|$.

The eigenvalues and eigenfunctions of the Hamiltonian can be expressed as

$$E_- = \frac{\hbar\sqrt{\Omega_1^2(t) + \Omega_2^2(t)}}{2}, \quad (44a)$$

$$|u_-\rangle = \frac{1}{\sqrt{2}}(\sin\Theta|0_L\rangle - |em\rangle + \cos\Theta|0_R\rangle) \quad (44b)$$

$$E_0 = 0, \quad (44c)$$

$$|u_0\rangle = \cos\Theta|0_L\rangle - \sin\Theta|0_R\rangle, \quad (44d)$$

$$E_+ = -\frac{\hbar\sqrt{\Omega_1^2(t) + \Omega_2^2(t)}}{2}, \quad (44e)$$

and

$$|u_+\rangle = \frac{1}{\sqrt{2}}(\sin\Theta|0_L\rangle + |em\rangle + \cos\Theta|0_R\rangle), \quad (44f)$$

with

$$\Theta(t) = \tan^{-1}\frac{\Omega_1(t)}{\Omega_2(t)}. \quad (45)$$

Here, time-dependent Rabi frequencies, $\Omega_1(t)$ and $\Omega_2(t)$, are defined as

$$\Omega_1(t) = \frac{\mu_{0,em}\cos\eta_L(t)A(t)}{\hbar}, \quad (46a)$$

and

$$\Omega_2(t) = \frac{\mu_{0,em}\cos\eta_R(t)A(t)}{\hbar}, \quad \text{respectively.} \quad (46b)$$

The above expressions have the same structures as those of a conventional STIRAP. It can be seen from Eq. (46) that a complete population transfer can be achieved by taking into account only the adiabatic change in the polarization direction of the linearly polarized electric field.

The population of $|0_R\rangle$, $P_{0R}(t)$, due to the transfer from $|0_L\rangle$ is expressed as

$$P_{0R}(t) = \sin^2 \Theta(t), \quad (47a)$$

where

$$\Theta(t) = \tan^{-1} \frac{\cos \eta_L(t)}{\cos \eta_R(t)}. \quad (47b)$$

The complete population transfer can be performed by setting the adiabatic changes both in $\eta_L(t)$ and $\eta_R(t)$ from the initial time $t = 0$ to the final time $t = t_f$ as

$$\eta_L(0) = \frac{\pi}{2} \quad \text{and} \quad \eta_R(0) \neq \frac{\pi}{2}; \quad \eta_L(t_f) \neq \frac{\pi}{2} \quad \text{and} \quad \eta_R(t_f) = \frac{\pi}{2}. \quad (48)$$

Let us define the relative angle between two transition moment vectors $\mu_{0L,em}$ and $\mu_{0R,em}$ as α $(0 \leq \alpha \leq \pi)$

$$\alpha = \eta_L(t) - \eta_R(t). \quad (49)$$

Figure 8 shows a 2D picture of the time-dependent $\eta(t)$. A complete population transfer from $|0_L\rangle$ to $|0_R\rangle$ is obtained by letting $\eta_R(t)$ be swept as

$$\eta_R(0) = \frac{\pi}{2} - \alpha \rightarrow \eta_R(t_f) = \frac{\pi}{2} \quad \left(\text{for } 0 \leq \alpha \leq \frac{\pi}{2}\right), \quad (50a)$$

and

$$\eta_R(0) = \frac{3\pi}{2} - \alpha \rightarrow \eta_R(t_f) = \frac{\pi}{2} \quad \left(\text{for } \frac{\pi}{2} \leq \alpha \leq \pi\right) \quad (50b)$$

Figure 9 shows the results of application of the new STIRAP to control of molecular chirality of a preoriented H$_2$POSH in a pure state case. A three-state model with the initial state $|0_L\rangle$ to the final state $|0_R\rangle$ and the fourth vibrational

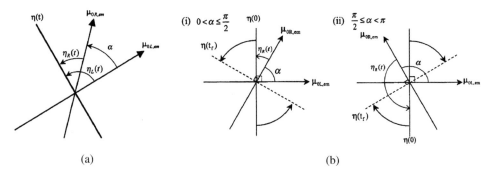

(a)   (b)

Fig. 8. (a) Time-dependent polarization vector $\eta(t)$, where α is the angle between two transition moments, $\mu_{0L,em}$ and $\mu_{0R,em}$ ($0 \leq \alpha \leq \pi$). $\eta_L(t)$ ($\eta_R(t)$) is the relative angle between $\eta(t)$ and $\mu_{0L,em}$ ($\mu_{0R,em}$). (b) Possible direction of the photon polarization vector $\eta(t)$ for $|0L\rangle \rightarrow |0R\rangle$: (i) $0 \leq \alpha \leq \pi/2$ and (ii) $\pi/2 \leq \alpha \leq \pi$. (Reprinted with permission from Ohta et al.[10] Copyright (2002) by American Institute of Physics).

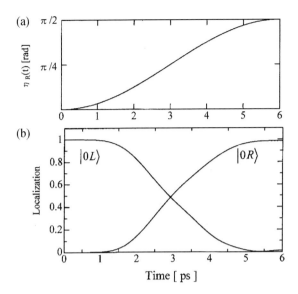

Fig. 9. Stimulated Raman adiabatic passage (STIRAP) applied to molecular chirality transformation from $|0_L\rangle$ to $|0_R\rangle$ of H_2POSH. (a) Form of the STIRAP variable $\eta_R(t)$ and (b) time-dependent populations of the two states, $|0_L\rangle$ and $|0_R\rangle$. (Reprinted with permission from Ohta et al.[10] Copyright (2002) by American Institute of Physics).

excited state in the first electronic excited state, $|e5\rangle$, was employed. Here, the electric field of the laser pulse applied has the form $\boldsymbol{E}(t) = \boldsymbol{\eta}(t) A_0 \cos \omega t$, in which $A_0 = 2.2 \times 10^9$ V/m,

$$\eta_L(t) = \frac{\pi}{2} + \frac{\pi}{2} \sin^2\left(\frac{\pi\, t}{2\, t_f}\right) \quad \text{and}$$

$$\eta_R(t) = \eta_L(t) - \frac{\pi}{2} \quad \text{with } t_f = 6.0\,\text{ps}.$$

It can be seen from Fig. 9 that the population is nearly completely transferred within the final time. The robustness of the STIRAP with respect to form of $\eta_L(t)$ ($\eta_R(t)$) and parameters, A_0 and t_f were confirmed.

In this section, a new STIRAP method for control of molecular chirality in the pure state case has been presented. The STIRAP method is also applicable to control of molecular chirality in the mixed case.[10]

4. Control of Helical Chirality

In this section, the results of quantum control of a chiral molecule possessing helical chirality are presented.[11] This is the first step for controlling the transformation between M and P forms of DNA. Difluorobenzo[c]phenanthrene is treated as a real helical system (Fig. 10).

Fig. 10. *M*- and *P*-forms of difluorobenzo[*c*]phenanthrene. Two yellow-colored balls in each form attached at the carbon denote chlorine atoms.

There are two fundamental issues regarding polyatomic molecules possessing helical chirality. The first issue is related to reaction coordinates. Generally, the most probable reaction path cannot be described in terms of a 1D symmetric or asymmetric double-well potential, because a symmetric structure is not always a true transition state for chirality transformation. An asymmetric transition-state structure should be described by more than one vibrational degree of freedom along reaction paths. Thus, it is essential to take multi-dimensional reaction coordinates into account.

The second issue is related to the mechanism of chiral transformation reactions. The tunneling mechanism works well for reactions with light atoms on its double-well potential. However, such reactions in a large molecule possessing helical chirality are practically forbidden in usual thermal conditions without any catalysis. Furthermore, there are two representative points in chiral transformation: one corresponds to the *M*-form and the other corresponds to the *P*-form. Therefore, it is necessary to treat these two representative points that move on its potential energy surface. These two issues are taken into account in this section.

The intrinsic reaction coordinate (IRC) or minimum energy path between *M*- and *P*-forms was generated by using the GAMESS suites of program codes.[12] The results of IRC analysis indicated that wagging motion (Q_1) of the CF$_2$ group and twisting motion (Q_2) of the benzene rings were principal modes for the transformation. Therefore, a 2D potential energy surface (Fig. 11) was employed. Two fluorine atoms were introduced for the dipole moment to be changed drastically along the reaction path.

A quantum control theory in a classical way was applied to the helical transformation in order to design laser fields.[13] Two representative points corresponding to the *M*- and *P*-forms were separately obtained by solving the time-dependent Schrödinger equation. This is a simplified treatment of quantum control of the chiral transformation in a racemic mixture. The amplitude of the control pulse is proportional to the linear momentum of the reaction system within the dipole approximation for the system-radiation field interaction.[13] The essential point for the control procedure is that the kinetic energy of the system is the controlling parameter. That

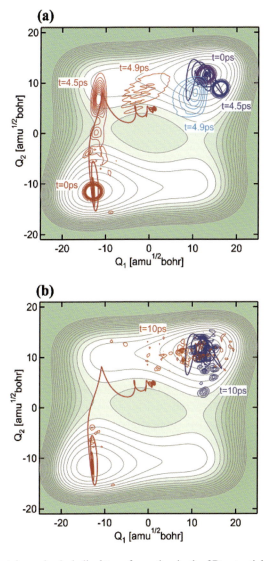

Fig. 11. Wavepacket dynamics in helical transformation in the 2D potential model. The thick red and blue lines denote the mean trajectories of the M- and P-forms, respectively. The contours with the series of reddish colors show time evolution of the M-form $|\Psi^{(M)}(t)|^2$, while those with the series of bluish colors show time evolution of the P-form $|\Psi^{(P)}(t)|^2$. (a) $t = 0 - 4.9$ ps. (b) $t = 10$ ps (end of the control period). (Reprinted with permission from Umeda et al.[11] Copyright (2002) by American Chemical Society).

is, the reaction is controlled by accelerating the representative point on a potential energy surface before crossing over a potential barrier and then by deaccelerating it to the target after passing over the potential barrier. The classical treatment was extended to control of wavepacket dynamics by replacing the classical momentum by a quantum-mechanically averaged momentum on the basis of the Ehrenfest theorem.

Difluorobenzo[c]phenanthrene was assumed to be preoriented at a space-fixed position. A preoriented difluorobenzo[c]phenanthrene from the M-form to P-form was controlled by two linearly (x- and z-) polarized IR pulses, $E_x(t)$ and $E_z(t)$.

The M- and P-forms are expressed in terms of two corresponding representative points, respectively. To control the two representative points at the same time in the molecular chiral exchange reaction, we adopted a simplified treatment based on the time-dependent Schrödinger equation.

In this simplified treatment, first, we individually evaluated the electric field of laser pulses $E^{(M)}(t)$ and $E^{(P)}(t)$, which are constructed from information on the wavepackets of the M- and P-forms, respectively. That is, the time-dependent Schrödinger equations are solved for the M- and P-forms separately. This means that this simplified treatment is a quantum control of molecular chirality in a racemic mixture. These two fields were averaged to obtain the total control field, that is, $E(t) = aE^{(M)}(t) + bE^{(P)}(t)$, where a and b are constants.[11]

Figures 11(a) and 11(b) show the temporal behaviors of the transformation in the presence of the control field. The red and blue lines are the mean trajectories of the M- and P-forms, respectively. It can be seen that the representative point initially representing the M-form moves to the target region of the P-form after crossing over the reaction barrier along the IRC, while that of the P-form remains in the original potential well. To visualize the dynamics of the isomerization process, the nuclear wavepacket $|\Psi(t)|^2$ at 0, 4.5, and 4.9 ps is depicted by contours in Fig. 11(a) and that at $t = 10$ ps is depicted in Fig. 11(b). The wavepacket of the M-form $|\Psi^{(M)}(t)|^2$ at $t = 4.5$ ps, that is, just after crossing over the reaction barrier, still retains a well-localized structure with slight delocalization. The delocalized part is because of a reflection by the barrier near the transition state. The delocalization of the wavepacket at 4.9 ps is because of the fact that the packet passes the flat area along the IRC.

The wavepacket at 10 ps is also delocalized widely in the potential well of the P-form. This indicates that hot enantiomers of the P-form are created. Wavepackets originally representing the enantiomer of the P-form $|\Psi^{(P)}(t)|^2$ are delocalized within its original potential well even in the presence of control pulses. The isomerization from the M-form to P-form is completed with a yield of ca. 70% within 3–4 ps, and the inverse process is suppressed by the control-field condition.

Figure 12 shows the amplitudes of laser fields E_x and E_z designed by the control method. This figure indicates that they are π-phase-shifted from the initial time to around 4.5 ps, while they are not phase-shifted after 4.5 ps. These phase behaviors are essential in controlling isomerization reactions such as molecular chiral reactions by using linearly polarized laser fields.

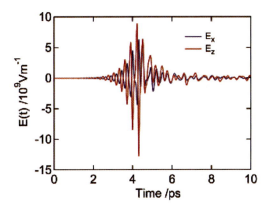

Fig. 12. Designed control fields. E_x (blue) and E_z (red) denote the x- and z-polarized IR pulses, respectively. They are π-phase-shifted at $t = 0 - 4.5$ ps. After the representative point of the M-form runs across the energy barrier, the phase of E_x is changed ($t > 4.5$ ps). (Reprinted with permission from Umeda et al.[11] Copyright (2002) by American Chemical Society).

5. Quantum Control in a Randomly Oriented Racemic Mixture Using Three Polarization Components of the Electric Fields

So far chiral molecules have been assumed to be oriented or attached to a surface. In this section, a quantum control of chiral molecules in a randomly oriented racemic mixture is presented.[14] As already introduced in Sec. 3 of Chap. 1 quantum control methods have been applied to chemical reaction dynamics in gases and also in solution.[15] In this section, the control is based on utilization of only the electric field components of laser pulses, not the magnetic field component. The control scenario consists of two kinds of lasers: a UV pulse laser and an IR laser. The IR laser pulse is prepared as a tool for orienting molecules in a homogeneously distributed system. UV lasers produce a sufficient enantiomeric excess in an ultra-short time scale when an oriented racemic mixture is prepared.

Let the electric fields of three pulsed lasers be $E_X(t)$, $E_Y(t)$, and $E_Z(t)$. Here, suffix X, Y, and Z of $E(t)$ denote the polarization directions of the electric fields in the space fixed coordinates. One laser, $E_Z(t)$, is used for orienting enantiomers, and the other two, $E_X(t)$ and $E_Y(t)$, are used for preparing pure enantiomers from preoriented ones.

Let us take H_2POSH as a realistic model again to demonstrate the control scenario. Figure 13 shows a 3D model in which the system consists of rotational degrees of freedom (θ and ψ) the P–S stretching vibrational mode R, and reaction coordinate ϕ that is the torsion of the SH bond around the PS bond. The rotation of H_2POSH was assumed to be a quasi-diatomic. The stretching mode was assumed to be a harmonic

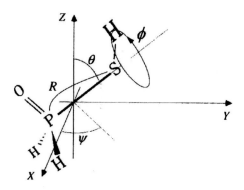

Fig. 13. An arrangement of H$_2$POSH in space fixed Cartesian coordinates (X, Y, Z). Angles θ and ψ denote the rotational variables of H$_2$POSH in a quasi-diatomic model, R denotes the vibrational mode of the PS bond, and ϕ denotes the coordinate of chirality transformation.

one. The rotational and vibrational constants were evaluated using an *ab initio* MO method.

The total Hamiltonian $H(t)$ is expressed as

$$H(t) = H_0 + V_Z(t) + V_X(t) + V_Y(t). \tag{51}$$

Here, the zero-order Hamiltonian H_0 is given as a sum of rotational Hamiltonian $H_0(\theta, \psi)$, vibrational Hamiltonian $H_0(R)$, and chirality-transformation Hamiltonian $H_0(\phi)$ as

$$H_0 = H_0(\theta, \psi) + H_0(R) + H_0(\phi). \tag{52}$$

In Eq. (51), $V_Z(t)$ is the interaction Hamiltonian between H$_2$POSH and the IR laser, and $V_X(t)$ and $V_Y(t)$ are the interaction Hamiltonians between H$_2$POSH and UV laser pulses.

These are expressed as[16]

$$V_Z(t) = -\mu_Z(R)E_Z(t)\cos\theta - \frac{1}{2}[\alpha_\parallel(R)\cos^2\theta + \alpha_\perp(R)\sin^2\theta]E_Z^2(t), \tag{53a}$$

$$V_X(t) = -\mu_X(\phi)E_X(t), \tag{53b}$$

and

$$V_Y(t) = -\mu_Y(\phi)E_Y(t). \tag{53c}$$

Here, $\mu_Z(R)$, $\mu_X(R)$, and $\mu_Y(R)$ are Z, X, and Y components of the dipole moment vector, respectively, and $\alpha_\parallel(R)$ and $\alpha_\perp(R)$ are parallel and perpendicular components of the polarizability, respectively.

Now look at control of orientation of randomly distributed H$_2$POSH. Here, $\cos\theta (0 \leq \theta \leq \pi)$ is taken as an operator for a measure of orientation. The orientation control is carried out by maximizing $\text{Tr}\{\rho(t_f)\cos\theta\}$ under the condition of a minimum input of the IR laser. The optimal laser pulse can be designed by maximizing the objective functional $O(E_Z)$ given as

$$O(E_Z) = \text{Tr}\{\rho(t_f)\cos\theta\} - \frac{1}{\hbar A}\int_{t_0}^{t_f} dt\, E_Z(t)^4, \quad (54)$$

where $\rho(t_f)$ is the density operator at final time t_f, and A is a weight-factor. The fourth power of $E_Z(t)$ instead of the square power is adopted for a fast convergence of the optimal procedure.[14]

Figure 14 shows the results of quantum control of orientation of H_2POSH starting from a randomly oriented racemic mixture.[14] In Fig. 14(a), the temporal behavior of $\langle\cos\theta\rangle$ is shown as a measure of magnitude of the orientation. The orientation was set to be carried out within 1 ps. It can be seen from Fig. 14(a) that the orientation is completed at the final time. Figure 14(b) shows the amplitudes of the optimal laser pulse for the orientation. The inserted figure in Fig. 14(b) shows the frequency-resolved spectrum of the optimal pulse: the strongest intensity band around $1,400\,\text{cm}^{-1}$ corresponds to the fundamental frequency of the P–S vibration. The coherent excitation of the fundamental and overtone vibrational bands of the P–S bond is the main origin for creation of the orientation. These two bands originate, respectively, from dipole and polarizability interactions terms in Eq. (53a). The former is of g-u transition and the latter is of g-g transition from the symmetry argument. The interaction between these two transitions creates asymmetry of the system, $i.e.$, the orientation shown in Fig. 14 (a).

Another oriented configuration, which is opposite to that derived above, can be prepared by applying the laser pulse with the out-of-phase $-E_Z(t)$.[14]

Let us produce L-enantiomers from a preoriented H_2POSH racemic mixture by using linearly polarized UV lasers. For simplicity, $H_0(\phi)$ was assumed to be expressed within a five-state model as

$$H_0(\phi) = |0_R\rangle\varepsilon_0\langle 0_R| + |0_L\rangle\varepsilon_0\langle 0_L| + |1_R\rangle\varepsilon_1\langle 1_R| + |1_L\rangle\varepsilon_1\langle 1_L| + |e\rangle\varepsilon_e\langle e|, \quad (55)$$

where $|n_R\rangle$ and $|n_L\rangle$ denote R- and L-enantiomer states with torsional vibrational level n and $|e\rangle$ denotes a vibronic eigenstate of the electronic excited state. ε_0, ε_1, and ε_e are the energies. The initial density operator $\rho(t_0)$ is expressed in the low temperature limit as

$$\rho(t_0) = |0_R\rangle\frac{1}{2}\langle 0_R| + |0_L\rangle\frac{1}{2}\langle 0_L|. \quad (56)$$

The orientation-controlled H_2POSH still has freedom of rotation around the Z-axis. We pick up two rotation-fixed configurations; one with $X, Y,$ and Z and the other with $-X, -Y,$ and Z, $i.e.$, the π-rotated one. Chirality transformation of L-enantiomers in the racemic mixture with the two configurations can be carried out by applying

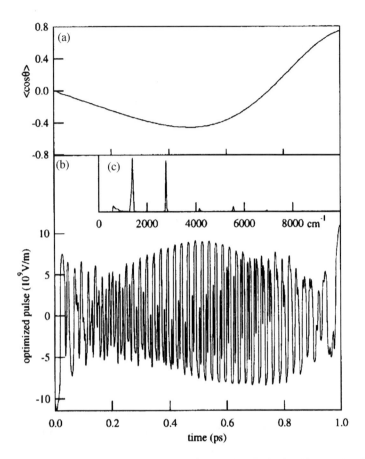

Fig. 14. (a) Temporal behaviors of the orientation of H$_2$POSH. (b) Optimal laser pulse shape for the orientation. (c) The inserted figure shows the frequency-resolved spectrum. (Reprinted with permission from Hoki et al.[14] Copyright (2002) by American Institute of Physics).

the pump–dump method described in a previous section: first a laser polarization direction of the pump laser pulse is chosen as

$$\langle 0_L | (\pm \mu_X, \pm \mu_Y) \cdot \begin{pmatrix} E_X(t) \\ E_Y(t) \end{pmatrix} | e \rangle \neq 0, \quad (57a)$$

and

$$\langle 0_R | (\pm \mu_X, \pm \mu_Y) \cdot \begin{pmatrix} E_X(t) \\ E_Y(t) \end{pmatrix} | e \rangle = 0. \quad (57b)$$

Here, signs ± correspond to the two configurations.

Figure 15 shows the results of temporal behaviors of the H$_2$POSH populations produced by a pump–dump pulse control.[14] Intensities of the pump and dump pulses were set to be 1.10×10^{10} V/m and 4.0×10^9 V/m, respectively. The pulse duration

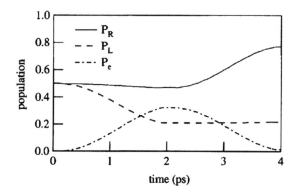

Fig. 15. Temporal behaviors of the populations of H_2POSH. A line denotes the population of the R-handed enantiomers, broken line that of the L-enantiomers, and dotted broken line that of the electronic excited state. (Reprinted with permission from Hoki et al.[14] Copyright (2002) by American Institute of Physics).

was taken to be 2 ps for both the pulses, which is long enough compared with a rotational period of $\sim$16 ps. The directions of the polarization were set at angle of $6°$ and $-6°$ degree with respect to the X-axis for the pump and dump processes, respectively. A high yield of $P_R(t_f) = 0.77$ was obtained at $t_f = 2$ ps. The results of simulation indicate that the chirality transformation in a randomly oriented racemic mixture can be performed by using three polarization components of electric fields of lasers.

In this chapter, first a basic principle for quantum control of molecular chirality transformation of preoriented enantiomers in a racemic mixture is described. Here, only linearly polarized electric field components of laser are utilized. Second, model calculations of transformation of enantiomers with axial chirality by using a pump–dump pulse method and a stimulated Raman adiabatic passage method have been carried out. A quantum control theory in a classical way, which is suitable to multi-mode systems, was applied to the transformation of enantiomers with helical chirality. Finally, a molecular chirality transformation in a randomly oriented racemic mixture using three polarization components of electric fields was theoretically demonstrated. Laser control of chiral molecules is a fascinating experimental target and will be realized in near future from the fact of significant development of laser-orientation control described in Chap. 4 and advances in laser technology.

References

1. (a) J. J. Flores, W. A. Bonner and G. A. Massey, *J. Am. Chem. Soc.* **99**, 3622 (1977), references therein; (b) Y. Inoue, *Chem. Rev.* **92**, 741 (1992).

2. (a) J. R. Cronin and S. Pizzarello, *Science* **275**, 951 (1997); (b) J. Bailey, A. Chrysostomou, J. H. Hough, T. M. Gledhill, A. McCall, S. Clark, F. Ménard and M. Tamura, *Science* **281**, 672 (1998); (c) Y. Inoue, H. Tsuneishi, T. Hakushi, K. Yagi, K. Awazu and H. Onuki, *Chem. Commun.* 2627 (1996); (d) S. Pizzarello and J. R. Cronin, *Geochim. Cosmochim. Acta* **64**, 329 (2000).
3. (a) M. Shapiro and P. Brumer, *J. Chem. Phys.* **95**, 8658 (1991); (b) J. A. Cina and R. A. Harris, *J. Chem. Phys.* **100**, 2531 (1994); *Science* **267**, 832 (1995); (c) J. Shao and P. Hänggi, *J. Chem. Phys.* **107**, 9935 (1997); *Phys. Rev. A* **56**, R4397 (1997); (d) A. Salam and W. J. Meath, *J. Chem. Phys.* **106**, 7865 (1997); *Chem. Phys.* **228**, 115 (1998); (e) Y. Fujimura, L. González, K. Hoki, J. Manz and Y. Ohtsuki, *Chem. Phys. Lett.* **306**, 1 (1999); *ibid.*, **310**, 578 (1999); (f) M. Shapiro, E. Frishman and P. Brumer, *Phys. Rev. Lett.* **84**, 1669 (2000); (g) Y. Fujimura, L. González, K. Hoki, D. Kröner, J. Manz and Y. Ohtsuki, *Angew. Chem. Int. Ed.* **39**, 4586 (2000), *Angew. Chem.* **112**, 4785 (2000); (h) R. P. Duarte-Zamorano and V. Romero-Rochín, *J. Chem. Phys.* **114**, 9276 (2001).
4. Y. Fujimura, L. González, K. Hoki, J. Manz, Y. Ohtsuki and H. Umeda, *Adv. in Multiphoton Processes and Spectroscopy*, Vol. 14 (World Scientific, Singapore, 2001), p. 30.
5. K. Hoki and Y. Fujimura, in A. D. Bandrauk, Y. Fujimura and R. J. Gordon (Eds.), ACS Books *Laser Control and Manipulation of Molecules* (American Chemical Society, 2002), p. 32.
6. K. Hoki, L. González and Y. Fujimura, *J. Chem. Phys.* **116**, 2433 (2002).
7. L. González, D. Kröner and I. R. Solá, *J. Chem. Phys.* **115**, 2519 (2001).
8. K. Hoki, L. González, M. F. Shibl and Y. Fujimura, *J. Phys. Chem. A* **108**, 6455 (2004).
9. (a) J. R. Kuklinski, U. Gaubatz, F. T. Hioe and K. Bergmann, *Phys. Rev. A* **40**, 6741 (1989); (b) U. Gaubatz, P. Rudecki, S. Schiemann and K. Bergmann, *J. Chem. Phys.* **92**, 5363 (1990); (c) K. Bergmann, T. Theuer and B. W. Shore, *Rev. Mod. Phys.* **70**, 1003 (1998).
10. Y. Ohta, K. Hoki and Y. Fujimura, *J. Chem. Phys.* **116**, 7509 (2002).
11. H. Umeda, M. Takagi, S. Yamada, S. Koseki and Y. Fujimura, *J. Am. Chem. Soc.* **124**, 9265 (2002).
12. M. W. Schmidt, K. K. Baldridge, J. A. Boatz, S. T. Elbert, M. S. Gordon, J. H. Jensen, S. Koseki, N. Matsunaga, K. A. Nguyen, S. Su, T. L. Windus, M. Dupuis and J. A. Montgomery Jr. *J. Comput. Chem.* **14**, 1347 (1993).
13. (a) H. Umeda and Y. Fujimura, *J. Chem. Phys.* **113**, 3510 (2000); (b) *Chem. Phys.* **274**, 231 (2001).
14. K. Hoki, L. González and Y. Fujimura, *J. Chem. Phys.* **116**, 8799 (2002).
15. (a) C. M. Dion, A. D. Bandrauk, O. Atabek, A. Keller, H. Umeda and Y. Fujimura, *Chem. Phys. Lett.* **302**, 215 (1999); (b) A. Keller, C. M. Dion and O. Atabek, *Phys. Rev.* **A61**, 23049 (2000); (c) K. Hoki and Y. Fujimura, *Chem. Phys.* **267**, 187 (2001).
16. (a) P. W. Brumer and M. Shapiro, *Principles of the Quantum Control of Molecular Processes*, (John Wiley and Sons, 2003); (b) *Laser Control of Quantum Dynamics*, Special Issue of *Chem. Phys.* Vol. 267 (2001); (c) D. J. Tannor and S. A. Rice, *Adv. Chem. Phys.* **70**, 441 (1988).

Index

13-cis retinal, 137, 139
1D control, 17
1D double-well, 148, 154
2,5-dichloro[n](3,6)pyrazinophane, 123, 127
$3.17 U_p$, 9
3D control, 17
3D model, 45, 167

above-threshold dissociation, 1, 12
above-threshold ionization, 1, 11, 55
acetaldehyde, 114
acetone, 114
achiral, 131
acoustic modulator, 4
ADI method, 47
adiabatic, 1, 3, 17, 28, 29, 58–62, 71, 73, 78, 81, 85–87, 89–92, 100, 102–104, 111, 121, 127–130, 142, 143, 147, 159, 162, 163, 171
adiabatic approximation, 87, 90
adiabatic parameter, 81, 85
adiabatic regime, 28, 58, 59, 61, 62, 73
adiabaticity parameter, 89, 93
ADK (Ammosov-Delone-Krainov) theory, 19
all-optical molecular orientation, 60, 62
allene, 114
alternating-direction implicit method, 43, 45
Ammosov-Delone-Krainov (ADK) model, 19, 64
angular momentum, 9, 46, 47, 61, 83, 96, 117, 122, 123, 125, 127–130
angular quantum numbers, 19
anisotropic hyperpolarizability interaction, 61
anisotropic polarizability, 16, 28, 56, 61, 62, 77
anisotropic polarizability interaction, 16, 28, 56, 61, 62, 77
aromatic ring, 123, 127, 131
asymmetric destruction, 147

asymmetric photolysis, 147
ATD, 1, 12, 13
ATD mechanism, 12
ATD spectrum of a hydrogen molecule, 12
ATI of molecular hydrogen, 11
atomic potential barrier, 14
atomic units, 7, 10, 19, 46, 52, 68, 96
atomic wavefunction, 19
attosecond pulse generation, 27
attosecond streak camera, 75
axial chirality, 147, 152, 171

backward fragment, 60
bacteriorhodopsin (bR), 133, 134, 137
bandwidth, 124, 158
barrier, 7, 14, 27, 72, 83, 95, 148, 152, 165–167
barrier height, 14, 152
barrier suppression ionization, 95
bending angle, 71, 72, 99, 100
benzene, 81, 104–108, 114, 117, 120–122, 124, 164
biochemistry, 17, 72
Born-Oppenheimer approximation, 3, 37, 86, 87, 128
Born-Oppenheimer electronic states, 84
bound state, 8, 50, 51
boundary, 41, 42
boundary conditions, 42
branching ratio, 67
breakdown of the adiabatic approximation, 90
breakdown of the Born-Oppenheimer approximation, 86, 87
brute-force orientation, 16
Buckminster fullerene, 109, 112
butadiene, 114

C_2H_2, 103
C_2H_4, 103
$C_2H_5Fe(CO)_2Cl$, 5
$C_4H_i^+$ ($i = 2 - 4$), 105
$C_5H_3^+$, 105
C_{60}, 109–112
$C_6H_5^+$, 105
carbon dioxide, 81, 114
carrier frequency, 83, 149
carrier-envelope phase, 73
CASPT2, 118
Cayley form, 43
Cayley-Crank-Nicholson scheme, 43, 44
CCN, 43, 44
central frequency of laser, 121
CH_3OH, 113
charge resonance state, 84
charge transfer state, 111
chiral aromatic molecule, 117, 118, 122, 127, 131
chiral transformation, 18, 151, 159, 164
chirality, 1, 17–19, 71, 72, 118, 129, 131, 147–154, 158, 160, 162–164, 166, 168, 171
chirality transformation, 18, 19, 148–151, 153, 154, 160, 163, 171
chirped pulse amplification, 27, 29
CI method, 100
circularly polarized, 9, 31, 57, 65, 76, 77, 117–121, 125, 131, 147
circularly polarized laser field, 9, 117, 125
cis-stilbene, 133–136
cis-trans isomerization, 133–136, 139–141, 143
classical model, 10, 14, 15
classical treatment, 165
clockwise, 118, 119, 124, 126, 127, 131
CO_2, 99–103
CO_2^+, 100
CO_2^{2+}, 100, 102, 104
coherent control, 133
coherent excitation, 2, 122, 169
coherent motions of vibrational modes in solution, 143
coherent nonlinear optical process, 7
coherent phase control, 4
coherent superposition, 38, 39, 139
coincidence momentum imaging, 99
coincident momentum imaging (CMI), 114
collective dynamic polarization, 111
COLTRIMS (COLd Target Recoil-Ion Momentum Spectroscopy), 26, 65, 75

complete population transfer, 159, 161, 162
completeness, 38
conical intersection, 133, 134, 139–143
constructive, 4
continuum state, 12, 49, 50
contour map, 48
coordinate representation, 41
correlation effects, 20
Coulomb energy, 14
Coulomb explosion, 1, 13–15, 25, 59, 64, 99, 100, 103, 104, 113
Coulomb explosion imaging, 25, 59, 64
Coulomb law, 13, 14
Coulomb singular points, 45
Coulomb-assisted ionization, 65
counter-intuitively, 159
counterclockwise, 118, 124, 126, 127, 131
coupling mode, 127, 128, 141–143
coupling operator, 86, 87
covariance mapping technique, 104
covariant map, 99
critical bond distance, 104
critical distance, 15
critical geometry, 104
critical internuclear distance, 14, 15
crossover, 4
current-induced magnetic moment, 117
cutoff, 7–11
cutoff energy, 10, 11
cutoff law, 9, 10
cycle-averaged potential, 99, 102, 111, 112, 114
cycle-averaged potential energy, 99, 114
cyclic aromatic molecule, 104, 117
cyclohexane, 114
cyclopentadienyl-iron-dicarbonyl-chloride, 5
cylindrical coordinates, 46, 82, 84

D_2^+, 113
DCP, 127, 128
DCPH, 122, 123, 125–127
decomposition, 147
degenerate electronic state, 120, 122
delayed ionization, 110
delocalization, 166
delta function excitation, 37, 38
density operator, 140, 148, 149, 151, 155, 169
dephasing, 2, 143
destructive, 4, 70, 71
diamagnetic ring current, 120
diatomic molecule, 8, 13, 14, 20, 64, 70, 81, 103

dichloropyrazine, 127, 128
difluorobenzo[c]phenanthrene, 163, 164, 166
dipole interaction, 62, 83, 95, 102, 147, 149
dipole moment, 5, 28, 30, 31, 36, 56, 58, 68, 73, 77, 83, 123, 141, 147, 148, 153–155, 158, 161, 164, 168
dipole moment operator, 5, 36
dipole moment vector, 168
dipole selection rule, 7
dipole-allowed transition, 106
direct ionization, 97, 98
dissociation dynamics, 48, 102
dissociation energy, 102
dissociative ionization, 3, 5, 13, 94
DNA, 163
doorway, 99, 110
doorway state, 99
double ionization, 55, 65, 97, 98
double-well, 14, 15, 18, 56, 60, 62, 148, 152, 154, 164
double-well electron potential, 14
double-well potential, 18, 56, 60, 62, 152, 164
dressed state representation, 151
dual transformation technique, 45

effective Coulomb charge, 19
effective potential, 83, 102, 103, 120
Ehrenfest theorem, 165
electron de Broglie wave, 68
electron density, 89, 121
electron mass, 47
electron rotation induced by laser pulses, 117
electron transfer, 3, 87–89, 91, 93, 100
electron transfer probability, 91, 93
electron-electron repulsion energy, 96
electron-nuclei correlated motions, 3
electronic and nuclear dynamics, 21, 49, 52, 81, 93, 109, 114
electronic angular momentum, 47, 83, 117, 128, 129
electronic dynamics, 15, 35, 45, 51, 87, 97, 100
electronic kinetic operator, 35
electronic ring current, 117–120, 122
electronic stereodynamics, 55
electronic wavepacket, 1, 97, 123
electronic wavepacket dynamics of H_2, 97
electrostatic interactions, 35
elliptically polarized laser field, 17, 30–32, 59
ellipticity, 31, 66, 71
emission model, 110

enantiomer, 17–19, 71–73, 118, 124, 129, 147–152, 154, 157, 166, 167, 171
enantiomeric excess, 147, 167
energy-gain, 8
enhanced ionization, 29–31, 51, 83
envelope function, 97, 101, 102, 119, 149, 158
equation of motion of the density matrix, 151, 156
equilibrium internuclear distance, 13, 48
error, 40
ethane, 114
ethylene, 114, 122, 123, 127
Euler angles, 16, 17, 58
Even-Lavie type pulsed valve, 78
exchange-correlation potential, 121
excited-state manifold, 110
explicit scheme, 44
exponential operator, 41
extreme ultraviolet (XUV), 7, 26, 112

fast Fourier algorithm, 41
feedback, 4, 58, 66, 134, 137
feedback approach, 134, 137
feedback loop, 4
femotosecond chemistry, 1
femtosecond transient impulsive Raman spectroscopy, 134
femtosecond transition-state dynamics, 133
FFT, 41
field adiabatic, 3
field ionization, 20, 52
field strength, 14, 100, 102, 111, 121
field-free 1D, 17
field-free 3D, 17, 59
field-free condition, 78, 100
field-free eigenstates, 151
field-induced nonadiabatic coupling, 87
field-induced nonadiabatic transition, 71
final condition, 5, 141
finite difference method, 41, 42
first-order time-dependent perturbation theory, 21
focal point, 4
Fourier slice theorem, 68
Fourier transform, 41, 68, 158
fractional step, 43
fragment, 13, 14, 25, 29–32, 56, 57, 59–61, 65, 71, 98, 99, 102–107, 110, 112, 113, 120, 147
fragment appearance energy, 112

fragment ion, 13, 14, 25, 29, 31, 61, 65, 71, 105, 106, 110, 113
fragmentation, 76, 81, 104, 106, 108–110
Franck-Condon (FC) position, 135
Franck-Condon excitation, 142, 143
Franck-Condon principle, 143
Franck-Condon state, 139
Franck-Condon wavepacket, 142
free electron, 50, 51, 71
free propagator, 37
free-free transition, 11
frequency-resolved optical gated measurement, 138
FROG, 138, 139
frozen-chemical bond approximation, 121
frozen-nuclei model, 128
full-width-at-half-maximum, 138
fullerene, 81, 109, 110, 112, 114
fundamental vibrational band, 169
furan (C_4H_4O), 104
FWHM, 138, 139

GA, 4, 5
GA experiment, 4
generalized cylindrical coordinates, 46
genetic algorithm, 4, 5, 66, 134, 137
giant plasmon, 111
grating mirror, 4
grid method, 45, 52
grid point method, 51, 52
grid points, 42, 48, 49
grid scheme, 45
grids, 41

H_2, 11–13, 20, 95, 97–100
H_2POSH, 72, 152, 153, 158, 162, 163, 167–171
H_3^+, 113
half time-point, 42
half time-step, 42
half-cycle pulse, 60
half-wave plate, 31
Hamiltonian matrix, 42, 92, 149
He, 7, 57, 95
head-versus-tail, 15, 32, 55
helical chirality, 147, 163, 164, 171
Hermitian operator, 40
hexapole focuser, 16
hexapole focusing, 16, 56, 78
HHG, 1, 7–11, 26, 27, 112

high-order harmonic generation (HHG), 1, 7–10, 25–27, 55, 68, 71, 73
highest-occupied molecular orbital (HOMO), 68
homochirality, 147
homogeneously distributed system, 167
homonuclear diatomic molecule, 81
hot enantiomers, 166
hydrocarbons, 113, 114
hydrogen molecular ions, 12, 45, 81, 94, 113, 114
hydrogen-like atom, 19
hyperpolarizability, 61, 62, 102

I_2, 14
implicit, 43–45, 97
impulsive excitation, 111
impulsive Raman, 17, 58, 111, 129, 131, 133–137
impulsive Raman excitation, 111
impulsive Raman process, 17, 58, 133, 134
in-phase, 124
induced dipole, 16, 28, 56, 58
induced dipole moment, 16, 28, 56, 58
inhomogeneous electrostatic field, 16, 56
initial phase, 10
injection-seeded Nd:YAG laser, 29
inner barrier, 14, 83
inner barrier height, 14
inner potential, 14
instantaneous frequency profile, 139
instantaneous mode, 135
instantaneous potential, 102, 104, 111
instantaneous time-dependent potential, 100
intense-field many-body S-matrix theory, 49
interaction picture, 36, 161
interelectronic correlation, 81, 95, 97, 98
interference, 2–4, 69–71, 76, 90, 91
interference between eigenstates, 3
interference between two phase-adiabatic states, 90
intermediate state, 44, 45, 98, 158, 160
interwell electron transfer, 87–89, 93
intramolecular dynamics, 49, 50
intramolecular electronic dynamics, 35, 51
intrinsic reaction coordinate, 164
inverse Fourier transform, 41
inversion symmetry, 7, 26
iodine, 14, 67
ion core, 9–11, 14, 51, 89, 94, 95

ionization potential, 6–9, 11, 12, 14, 19, 64, 68, 77
ionization probability, 51, 73, 77
ionization rate, 6, 19–21, 49, 64, 68, 77
IRC, 164, 166
isomerization coordinate, 137

Keldysh parameter, 6, 7, 105, 109
Keldysh-Faisal-Reiss theory, 49
KFR theory, 49

L-enantiomer, 18, 150, 151, 154, 155, 157, 158, 169, 171
Landau–Zener formula, 92
Laplacian, 128
large-amplitude oscillation, 111
large-amplitude vibration, 130
laser-induced dissociative ionization, 93
laser-induced electron diffraction (LIED), 75
laser-induced electronic ring current, 120
laser-induced nonadiabatic coupling, 86, 87
laser-induced ring current, 118
law of conservation of momentum, 71
learning-loop optimal control, 66
left-handed, 18, 121
level splitting, 153
light modulator, 4
linear combination, 18, 38, 118, 127, 142
linearly polarized laser, 9, 17, 28, 29, 57–59, 62, 81, 82, 117, 118, 122, 124, 127, 131, 141, 147, 150, 152
linearly polarized laser field, 29, 59, 127
Liouville equation, 151
liquid crystal, 4
local optimal control, 72
localization, 2
localized basis set, 90, 92, 93, 150, 154
localized state, 1, 3, 18, 89, 149, 152, 158
long-range force, 41
low-frequency limit, 6

magnesium-porphyrin, 117
magnetic dipole moment, 147
magnetic moment, 117, 131
many-body problem, 49
mass spectra, 66, 105, 106, 108, 109, 113
mass-resolved momentum imaging (MRMI), 99, 114
MCSCF, 100
methane, 114

methanol, 113, 136, 137
M-form, 164–167
Mg-porphyrin, 117–121
minimum energy path, 164
mirror image, 18, 89, 123, 148, 154
molecular ADK (MO-ADK) model, 64
molecular ADK theory, 19
molecular alignment, 3, 16, 17, 25, 27, 28, 30, 55–59, 64, 65, 68, 69, 73, 76, 78
molecular chirality, 1, 19, 118, 129, 131, 147, 148, 158, 160, 162, 163, 166, 171
molecular Hamiltonian, 4, 35, 50, 141, 149, 155, 161
molecular handedness, 17
molecular imaging, 27
molecular orbital tomography, 55
molecular polarizability, 8
molecular-frame photoelectron angular distribution (MFPAD), 76
momentum space, 41
Mulliken population, 100
multi-charged carbon atoms, 104
multi-electron system, 95
multi-mode model, 143
multiphoton, 1, 6, 7, 11, 27, 62, 63, 66
multiphoton ionization, 6, 7, 27, 62, 63, 66
mutation, 4

Ne, 7
nonadiabatic, 1, 3, 5, 17, 27, 28, 56, 58, 60, 63, 64, 69, 71, 74, 75, 78, 86, 87, 91–93, 110, 111, 117, 118, 127–131, 133, 142
nonadiabatic coupling, 5, 86, 87, 117, 118, 127–129, 131
nonadiabatic coupling operator, 86
nonadiabatic effects, 3
nonadiabatic interaction, 3
nonadiabatic kinetic energy correction, 86
nonadiabatic photodissociation reaction, 75
nonadiabatic regime, 28, 58, 74
nonadiabatic transition, 71, 87, 91–93, 130, 142
nonadiabatic transition probability, 91, 93
nonadiabatic treatment, 3
noncommuting, 40
noncommuting operator, 40
nonhelical polarized laser, 118
nonpolar molecule, 16, 56
nonresonant laser field, 17, 56, 59, 60, 74
nonsequential double ionization, 55, 65, 97
nonsequential multiple ionization, 95

norm, 40, 41, 43, 46
norm of the wavefunction, 40, 46
normalization condition, 46
normalized difference, 75
nuclear kinetic energy, 3, 82, 86
nuclear kinetic operator, 35, 46
nuclear wavepacket, 1–3, 35, 37–39, 51, 52, 89, 93, 94, 102, 111, 128, 133–135, 143, 166
nucleus-nucleus interactions, 35

objective functional, 5
oblate form, 111
one-bond breaking, 102, 104
one-electron atoms, 6, 21
one-to-one statistical mixture, 149, 155
optical centrifuge, 57
optical control, 133
optical cycle-averaged potential, 99, 102
optical excitation, 37, 38, 150
optical Franck-Condon overlap integral, 38
optical nonlinearity, 112
optical polarizability, 112
optimal control, 1, 5, 58, 66, 72, 127, 139, 141–143
optimal pulse, 139, 141, 169
orientation, 3, 15–17, 25, 28–33, 55–64, 66, 70, 76–78, 140, 141, 168–170
oriented molecules, 15–17, 55, 64, 73, 76–78
oriented racemic mixture, 167, 169, 171
orthogonality, 37
oscillation period, 7, 128, 129, 135, 158
out-of-phase, 124, 126, 130, 139, 169
outer potential, 14, 15
overlap integral, 38
overtone vibrational band, 169

parallel and perpendicular components of the polarizability, 168
parallel transition, 67
Pariser-Parr-Pople (PPP) model, 123
parity-violating force, 19
parity-violating interaction, 19
Peaceman-Rachford method, 44
peak of the HHG signal, 11
pendular state, 56
permanent dipole, 16, 28, 30, 31, 55, 56, 61, 62, 73, 77
permanent dipole interaction, 62
permanent dipole moment, 16
perpendicular transition, 67

P-form, 164–166
phase adiabatic, 3
phase matching, 68
phase-adiabatic electronic eigenstates, 85
phase-adiabatic potential, 85, 86, 89
phase-adiabatic states, 81, 84, 85, 87, 89–91, 114
π pulse, 118, 124
π-electron, 117, 118, 120, 122–127, 129, 131
phosphinotioic acid, 152
photo-induced current, 122
photo-induced isomerization, 133
photochemical reaction, 3, 5
photodissociation, 13, 67, 75, 159
photodissociation of H_2^+, 13
photoelectron, 11, 12, 75–77, 95
photoelectron angular distribution (PAD), 75, 76
photoelectron spectra, 11, 12
photoisomerization, 133–135, 137, 141–143
photon angular momentum, 117, 122
photon polarization, 8, 36, 37, 117, 123, 148, 151, 153, 155, 162
photon polarization unit vector, 36
photon-dressed potential, 106
photon-polarization vector, 149
photoreaction, 139
phototaxis, 133
plane of symmetry, 17
plane wave, 68
plasma shutter, 60
plateau, 7, 8
Poisson distribution, 143
polarizability, 8, 16, 28, 30–32, 56, 58, 59, 61, 62, 73, 74, 77, 102, 112, 168, 169
polarization direction, 56, 65, 73, 118, 122, 124, 126, 129, 131, 150, 153, 155, 162, 167, 170
polarization energy, 102
polarization vector, 124, 128, 157, 158, 162
polyatomic molecule, 3, 8, 15, 21, 71, 81, 103, 104, 113, 127, 143, 164
pondermotive energy, 6–8, 11
population transfer, 110, 130, 159–162
post form, 50
PR method, 44
precursor, 100
preoriented enantiomers, 147, 148, 150, 154, 171
projection operator, 50
prolate form, 111

propanol, 114
proton dynamics, 81
proton mass, 47, 82
pulse envelope function, 101, 149
pulse shaping, 4
pump process, 155–157
pump-dump control, 122, 125, 147, 154, 158
pumping process, 150, 155, 158
pure enantiomers, 18, 73, 147, 167
pyrazine ($C_4H_4N_2$), 104
pyridine (C_5H_5N), 104
pyrrol (C_4H_4NH), 104

quantum beat interference, 76
quantum beats, 3, 39, 76
quantum control, 4, 19, 133, 134, 140, 143, 147, 163, 164, 167, 169, 171
quantum control of molecular chirality, 19, 147, 166
quantum control theory in a classical way, 164
quantum dynamical treatment, 15
quantum interference, 69, 70
quantum revival, 17
quantum yield, 137
quarter-wave plate, 31
quasi-degenerate, 122, 123, 127, 153
quasi-degenerate π-electronic excited states, 123
quasi-degenerate π-electronic excited states, 122

R-enantiomer, 18, 19, 123–126, 131, 148–151, 153–155
Rabi frequency, 152, 153, 155, 157
Rabi oscillation, 154
racemate, 71, 72
racemic mixture, 19, 147–151, 153–155, 157, 158, 164, 166, 167, 169, 171
randomly oriented, 15, 21, 30, 32, 147, 167, 169, 171
randomly oriented molecules, 15
rare gas, 7, 73, 74
reaction barrier, 166
reaction coordinate, 18, 135, 136, 140–142, 164, 167
reaction microscope, 26
real-time, 7, 133, 134
real-time measurement, 7
real-time observation, 133, 134
real-time structural evolution, 133

recombination, 8, 10, 11, 27, 69–71
recombination step, 8
reference Hamiltonian, 49, 50
regulation function, 5
rephasing, 2
representative point, 164–167
reproduction, 4
rescattering of electron, 1
resonance condition, 161
resonant multiphoton ionization, 11
retinal, 133, 134, 137–143
retinal in bacteriorhodopsin, 133, 134
revival, 2, 17, 58, 59, 71, 76, 130
right-handed, 18
RIMS (Recoil-Ion Momentum Spectroscopy), 26
ring current, 117–122
rotating wave approximation, 39, 152, 156, 161
rotational angular momentum, 117
rotational constant, 59, 61, 74
rotational Hamiltonian, 168
rotational quantum states, 16, 60, 76
rotational wavepacket, 1, 17, 58, 64, 71
RWA, 161
Rydberg, 1

S-enantiomer, 123, 124, 126, 131
scaled coordinates, 45, 46
scattering matrix, 35, 49, 50
Schrödinger equation, 5, 6, 36, 38, 39, 41, 42, 45–47, 49, 50, 82, 83, 97, 111, 118, 123, 141, 164, 166
second-order accurate, 43
semiclassical treatment, 35, 36, 42, 91, 149
semiclassical way, 35
sequence energy deposition, 111
sequential nonadiabatic, 110
sine-square pulse, 51
singly and doubly excited configuration interaction (SDCI), 123
singularity, 41, 42, 45
SLM, 4
S-matrix, 49, 51, 52
soft x-ray, 27
solvent-dependence in isomerization rates, 137
space-fixed coordinates, 158
spatial direction, 16, 58
spatial light modulator, 4
spectator coordinate, 135
spherical harmonics, 19

split operator, 40, 41
split operator method, 40, 41
squared angular momentum operator, 61
Stark shift, 77
stationary state, 18
statistical weight, 149, 155
stereochemistry, 17
stereodynamics, 16, 55
stilbene, 133–135, 143
stimulated Raman adiabatic passage, 147, 159, 163
STIRAP, 159–163
Stokes pulse, 159
structural deformation, 71
superconductivity, 112
symmetric top molecule, 16, 59
symmetrized split operator, 40, 41
symmetrized split operator method, 40
symmetrized Trotter formula, 40

tandem time-of-flight mass (TOF–MS) spectrometer, 105
target operator, 5, 141, 149, 155, 158
TDDFT, 120, 121
thermal ensemble, 63, 78
three-point finite difference, 42
three-step model, 10, 73
time dependent, 3, 5, 21, 36, 37, 39–42, 45–50, 66, 72, 73, 82, 83, 89, 97, 100, 101, 111, 118–121, 123, 135, 137, 141, 142, 151, 152, 158, 160–164, 166
time evolution, 2, 37, 39, 40, 47, 127, 151, 155, 156, 158, 165
time evolution operator, 37
time splitting, 43
time-dependent density functional theory (TDDFT), 120, 158
time-dependent electron potential, 3
time-dependent electronic current density, 118, 119
time-dependent Hartree potential, 121
time-dependent Kohn-Sham equation, 120
time-dependent Lagrange multiplier, 141
time-dependent polarization pulse, 66, 67, 72
time-dependent polarization vector, 160
time-dependent population, 141, 152, 163
time-dependent Schrödinger equation, 97
time-flight-spectrometer, 28
time-independent, 38, 40, 43, 111
time-of-flight (TOF) spectrometer, 28

time-ordering operator, 36
time-resolved, 2, 39, 75, 129, 135
time-resolved impulsive Raman, 135
TOF mass spectra, 105, 106, 108, 113
tomographic imaging, 68
torsional vibrational, 169
trans-cis isomerization, 133
transform-limited pulse, 138
transient absorption, 135
transient ring current, 117
transient thermal electron, 110
transition dipole moment, 68, 123, 158, 161
transition moment, 37, 38, 93, 127, 140, 155, 158, 162
transition state, 137, 164, 166
triatomic hydrogen molecular ions, 113
tridiagonal, 42, 44
tridiagonal system, 44
Trotter production formula, 40
tuning laser, 5
tunnel ionization, 64, 66, 68, 107
tunneling, 1, 6–8, 10, 18–21, 51, 60, 64, 75, 77, 95, 105, 109, 150–154, 158, 164
tunneling ionization, 1, 6–8, 19–21, 64, 75, 77, 95, 105
tunneling ionization rate, 19–21, 64, 77
tunneling mechanism, 164
tunneling time, 7, 153, 158
two point emitters, 70
two-bond breaking, 100, 102
two-dimensional (2D) ion imaging technique, 56
two-electronic state model, 1, 2, 37, 84
two-point formula, 42

unitary operator, 36, 40, 50, 155
unsaturated hydrocarbons, 133
UV, 7, 118, 147, 167–169

valence electron, 19, 64
valence orbital, 64, 71
van der Waals dimer, 73, 74
variational principle, 5
vector potential, 77
velocity map condition, 25
vibrational coherence, 135
vibrational eigenfunction, 38
vibrational wavepacket, 1, 111, 112, 142
vibronic states, 39
VIS, 7

vision, 133, 139
Volkov state, 6, 51

wavepacket, 1–4, 17, 35, 37–41, 48–52, 58, 64, 65, 68, 69, 71, 72, 89, 93–95, 97, 102–104, 111, 112, 117, 118, 123, 124, 127–131, 133–135, 139–143, 149, 158, 159, 165, 166
wavepacket methodology, 1

wavepacket propagation, 35, 41, 48, 49, 52, 127
Wigner transformation, 139

XUV, 7, 26, 27, 112

zero kinetic energy, 7
Zewail, 133